T0155870

Cell and Molecular Biology
for Non-Biologists

Lorenz Adlung

Cell and Molecular Biology for Non-Biologists

A Short Introduction into Key Biological Concepts

 Springer

Lorenz Adlung
University Medical Center Hamburg-
Eppendorf
Hamburg, Germany

ISBN 978-3-662-65355-5   ISBN 978-3-662-65357-9   (eBook)
https://doi.org/10.1007/978-3-662-65357-9

© Springer-Verlag GmbH Germany, part of Springer Nature 2022
This work is subject to copyright. All rights are reserved by the Publisher, whether the whole or part
of the material is concerned, specifically the rights of translation, reprinting, reuse of illustrations,
recitation, broadcasting, reproduction on microfilms or in any other physical way, and transmission or
information storage and retrieval, electronic adaptation, computer software, or by similar or dissimilar
methodology now known or hereafter developed.
The use of general descriptive names, registered names, trademarks, service marks, etc. in this publication
does not imply, even in the absence of a specific statement, that such names are exempt from the relevant
protective laws and regulations and therefore free for general use.
The publisher, the authors and the editors are safe to assume that the advice and information in this book
are believed to be true and accurate at the date of publication. Neither the publisher nor the authors or
the editors give a warranty, expressed or implied, with respect to the material contained herein or for any
errors or omissions that may have been made. The publisher remains neutral with regard to jurisdictional
claims in published maps and institutional affiliations.

Responsible Editor: Sarah Koch
This Springer imprint is published by the registered company Springer-Verlag GmbH, DE part of
Springer Nature.
The registered company address is: Heidelberger Platz 3, 14197 Berlin, Germany

Preface

This book reflects the diversity of Modern Biology with its plethora of emerging disciplines. It is dedicated to all non-biologists who introduce their knowledge and expertise to the field thereby enriching the scientific landscape in biology. Experts from all over the life sciences are needed to answer current questions in today's biology and medicine [99]. Facing the challenge to combat a disease as complex as cancer demands interdisciplinary collaboration.

Mathematics, computer sciences, physics and engineering became the vital core of disciplines such as systems biology and synthetic biology. A vivid exchange is a prerequisite for synergies and mutual benefits in the biological as well as the non-biological spheres. A major hurdle to overcome is language. Computational modelling allows a new understanding of non-intuitive biological phenomena. Concepts and formalisms are introduced for systematic scrutinization of biological systems and rational hypothesis testing. However, nomenclature and semantics are often very different among the fields, which leads to misconceptions that hamper communication and thus scientific progress. Once a common ground is discovered with a coherent language, findings can be shared to the full extent. This approach would enable an entirely new perspective on living entities that could then be described mechanistically and quantitatively.

This book addresses the fundamental principles of Modern Biology exemplified by the latest research results. The combination of both is described with a vocabulary that requires no pre-existing knowledge. Schemes are depicted in simple shapes boiled down to central motifs that could be re-drawn in any sketchbook. Many concepts were borrowed from foreign disciplines to highlight similarities and possible linking points for career changers. Wherever possible, open-access versions of original literature are cited. This book is intended as a brief introduction to Modern Biology for interested readers from other disciplines of natural sciences. It is recommended especially to people with a background in mathematics, computer sciences, physics or engineering, who are planning to study or work in biomedical life sciences. Please feel kindly encouraged to further foster a true interdisciplinary debate that facilitates scientific progress in Modern Biology.

Hamburg   Lorenz Adlung

Contents

Contents

Acronyms

4E-BP	4E-binding protein
ADP	adenosine diphosphate
AKT	thymoma viral proto-oncogene
APC	anaphase-promoting complex
APCs	antigen-presenting cells
ATM	adenine-thymine mutated
ATP	adenosine triphosphate
ATR	ATM- and Rad3-related
Bcl2	B cell lymphoma 2
BCR	B cell receptor
BRCA1	breast cancer type 1 susceptibility protein
cAMP	cyclic adenosine monophosphate
CD	clusters of differentiation
cdc	cell division cycle
CDK	cyclin-dependent kinase
CFTR	cystic fibrosis transmembrane conductance regulator
Chk	checkpoint kinase
CIS	cytokine-inducible SH2 domain-containing protein
CTD	C-terminal domain
CTLA-4	cytotoxic T-lymphocyte-associated protein 4
DAMPs	damage-associated molecular patterns
DNA	deoxyribonucleic acid
DNMT	DNA methyltransferase
DUSP	dual-specificity phosphatase
eIF-4E	eukaryotic initiation factor 4E
EMT	epithelial-mesenchymal transition
Epo	erythropoietin
EpoR	erythropoietin receptor
ER	endoplasmatic reticulum
ERAD	endoplasmatic reticulum-associated protein degradation
ERK	extracellular signal-regulated kinase
GAP	GTPase-activating protein
GDP	guanosine diphosphate
GEF	guanine nucleotide exchange factor

GFP	green fluorescent protein
GLUT1	glucose transporter 1
GTP	guanosine triphosphate
HAT	histone acetyltransferase
HDAC	histone deacetylase
HLA	human leukocyte antigen
HSC	hematopoietic stem cell
IFN	interferon
$InsP_3$	inositol 3,4,5 trisphoshpate
JAK2	Janus kinase 2
LDL	low-density lipoprotein
MAPK	mitogen-activated protein kinase
MAPKK	mitogen-activated protein kinase kinase
MAPKKK	mitogen-activated protein kinase kinase kinase
MBD	methyl-CpG-binding domain
MC1R	melanocortin 1 receptor
Mdm2	Mouse double minute 2 homolog
MEK	mitogen/extracellular signal-regulated kinase
MET	mesenchymal-epithelial transition
MHC	major histocompatibility complex
MPF	mitosis-promoting factor
MRN	Mre11, Rad50, Nbs1
mRNA	messenger ribonucleic acid
mTOR	mammalian target of rapamycin
NF2	Neurofibromin 2
NFκB	nuclear factor kappa-light-chain-enhancer of activated B cells
ORI	origin of replication
OST	oligosaccharyl transferase
PAMPs	pathogen-associated molecular patterns
PET	positron emission tomography
PI3K	phosphoinositide 3-kinase
PIP_2	phosphatidylinositol 3,4 bisphoshpate
PIP_3	phosphatidylinositol 3,4,5 trisphoshpate
PRRs	pattern-recognition receptors
PTEN	phosphatase and tensin homolog
Rb	retinoblastoma
RNA	ribonucleic acid
RPA	replication protein A
rRNA	ribosomal ribonucleic acid
SCF	Skp, Cullin, F-box containing
SNARE	soluble N-ethylamine sensitive factor attachment protein receptor
SNP	single-nucleotide polymorphism
SOCS3	suppressor of cytokine signalling 3
SRP	signal recognition particle
SSB	single-strand binding protein

STAT5	signal transducer and activator of transcription 5
TBP	TATA box binding protein
TCR	T cell receptor
TGF-β	transforming growth factor beta
TLRs	Toll-like receptors
TNF	tumor necrosis factor
TRAIL	TNF-related apoptosis-inducing ligand
tRNA	transfer ribonucleic acid
TSS	transcription start site
UPR	unfolded protein response
UV	ultra-violet

Cell Architecture

1

Contents

Introduction

What is a cell? A cell in biology is the fundamental unit of life [7]. All living entities are built from cells. Life in its essence means self-maintenance and replication.
And cells are nothing more than self-maintaining, replicating droplets. Replication
is performed by a sequence of biochemical reactions. To provide a separate compartment for those reactions to occur, cells are enclosed by a membrane surrounding
its aqueous inner solutions. To fulfil their function of self-maintenance and replication, different structures evolved inside the cells. Ensembles of molecules form
a machinery that builds up, maintains and replicates cellular contents. For replication, a blueprint of the cell has to be delivered as a construction plan for each newly
produced cell. This information is stored on deoxyribonucleic acid (DNA), a long
macromolecule. In sum, cells can be considered as a membrane-enclosed piece of
DNA in an aqueous solution (Fig. 1.1). Beside membrane and DNA, additional parts
exist to allow more sophisticated structures and thus more specialized functions to
be carried out inside the cells.

Our human body consists of a skeleton and organs. Likewise, the cellular body
contains a "cytoskeleton" and "organelles". The structural elements of a cell are its
building blocks and thus the bricks of cellular architecture. Diversity and complexity
of the structures vary depending on the cell's specialized function. While some cells

© Springer-Verlag GmbH Germany, part of Springer Nature 2022
L. Adlung, *Cell and Molecular Biology for Non-Biologists*,
https://doi.org/10.1007/978-3-662-65357-9_1

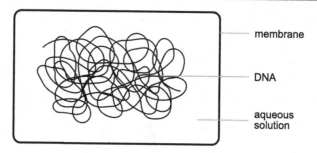

Fig. 1.1 A cell as central unit of life consists in its essence of a membrane that contains DNA as information storage in aqueous solution

remained rather simple like our red blood cells, others became sophisticated bioreactors full of specialized structural elements and derived functions like brain cells. From unicellular organism to higher life forms, the structure of biological entities is always tightly connected to its function. The emerging diversity that discriminates different species lays the foundation for life with all its variety.

1.1   Cells and Their Biochemical Composition

The components of living cells are diverse, but the majority of molecules belong to one of the four main categories, which are:

- carbohydrates ("sugars", "carbs"),
- lipids ("fats"),
- proteins and
- nucleic acids.

These macromolecules are assembled from building blocks: monosaccharides, fatty acids, amino acids and nucleotides (Table 1.1). These core molecules account for a large fraction of cellular mass. In fact, the cellular body should be disregarded as an aqueous solution because it is rather a dense gel-like mixture crowded with proteins made by ribosomes (see Sect. 1.3) [62]. Therefore, the term cytoplasm implies the correct constitution of the cellular body rather than the term cytosol.

Table 1.1 Monomers as building blocks for functionally distinct molecular assemblies

Bricks	Macromolecules	Function
Monosaccharides	Sugars	Energy source, regulatory units
Fatty acids	Fats	Membranes, energy storage
Amino acids	Proteins	Cell architecture, enzymes
Nucleotides	Nucleic acids	Information storage, regulatory units

Sugars are a main source of energy. Glucose plays a central role. It can be broken down for the release of energy, or it can be lined up in long chains for storage. Chains of monosaccharides can also be plugged onto the other molecule classes forming e.g. glycolipids or glycoproteins. Glycoproteins are also an integral part of the bacterial cell wall and the extracellular matrix in tissues of higher organisms.

Lipids consist of fatty acids and similar biomolecules such as waxes and vitamins. Fatty acids are important for membranes of cells (see Chap. 5) and their subunits ("organelles"). Breaking down fatty acids yields even more energy per mass unit than glucose. Lipids are stored in cytoplasmic droplets. Mostly, three fatty acids are connected to a glycerol molecule forming triacylglycerol, the main component of lipid droplets.

Proteins exert ubiquitous functions from shaping the cell's interior (see Sect. 1.2) to catalysing biochemical reactions as enzymes. The functional diversity originates from the modular structure. Proteins are long chains formed from 20 amino acids that serve as building blocks allowing for combinatorial complexity. Polymers of hundreds of amino acids fold into a 3D structure of the mature protein. Proteins are encoded by the sequence of nucleic acids in the genetic code (see Sect. 2.3).

Nucleic acids exist in two forms: ribonucleic acid (RNA) and deoxyribonucleic acid (DNA). The latter one is chemically more stable and used for long-term storage of hereditary information in the genetic code, which will be explained in the next chapter (see Chap. 2). RNA is less stable and carries short-term instructions as messenger molecule or regulatory subunit of protein machineries and cellular organelles (ribosomes).

As all these macromolecules are modularly formed from subunits, size scales are important to bear in mind (Table 1.2). Modularity is an efficient feature emerging from the evolution of life allowing almost unlimited complexity with a predefined set of building blocks that can be easily maintained and sustainably produced. However, it is not sufficient to know the mere sequence of subunits that form a protein or an RNA chain. The 3D structure of a molecule is essential for the exertion of its specific function. Structure and function are intimately related in biology. Enzymes can only catalyse biochemical reactions because they fit their substrates like "lock and key". A certain shape always carries a certain property, otherwise, it would be a waste of resources to maintain it in the course of evolution.

Built from these macromolecules at different scales, two kinds of cells can be discriminated.

1. **Prokaryotic cells**, derived from the Greek words *"pro"* and *"karyon"* meaning "before" and "kernel", refer to cells without an information-storing compartment. In prokaryotes, DNA as storage of heritable properties is not encapsulated by a membrane in a so-called nucleus. Instead, the DNA is loosely arranged within the

Table 1.2 Size scales of cellular components. $1\,nm = 10^{-9}\,m$, $1\,\mu m = 10^{-6}\,m$

Size	$0.1\,nm = 1\,Å$	$20\,nm$	$1.5\,\mu m$	$20\,\mu m$
Component	Atom	Molecule	Prokaryotic cell	Eukaryotic cell

cell throughout the entire aqueous solution called cytoplasm. Many cell-derived terms are commonly used with the prefix "cyto". Despite their rather simple and ancient structure (Fig. 1.1), prokaryotes show high diversity and abundance. Diversity originates from their ability to adapt to biospheres such as undersea volcanoes or our human guts. Abundance results from their ability to double quickly, like some gut bacteria within 20 min under optimal conditions. Robustness is provided by additional layers supporting the cellular membrane thereby creating a cell wall. Bacteria, the most prominent domain within prokaryotes, e.g. grows and evolves rapidly leading to the spread of disease and antibiotic resistance.

2. **Eurkaryotic cells** with the Greek prefix "*eu*" for "good" indicates that they do have a nucleus. Besides, they also contain many other specialized compartments called organelles (see Sect. 1.3). Some of these are derived from ingested, highly adapted bacteria that could produce energy from light or oxygen. Since maintenance of such organelles is time-consuming, eukaryotes grow slower but they are also bigger than the $\approx 1.5\,\mu m$ unicellular prokaryotes. Eukaryotes can exist as single cells, e.g. yeast, or live as multi-cellular assemblies creating all our body tissues.

1.2   The Cell's Skeleton

The cells of our body differ widely in their shape and thus in their function. The diversity of cellular structures can only be maintained by an underlying network of particles that comprise the cytoskeleton. These structures manifest tight cellular connections, e.g. the outer layer of our body: the skin. Unlike the skeleton of our body, the cellular skeleton, however, is a highly dynamic structure that can adapt in space and time to steer cellular re-organization, for instance, during the cell cycle (see Chap. 7).

The machinery of the cell's skeleton is based on three different types of protein filaments.

1. Actin filaments
   - ø: 5–9 nm, "micro filaments"
   - two intertwined helix polymers of actin, flexible (i.e. dynamic) meshes
2. Intermediate filaments
   - ø: 10 nm, heterogeneous
   - rope-like structure, stable
3. Microtubules
   - ø: 25 nm, basis (i.e. source) is centrosome
   - tube of Tubulin molecules, long, straight and compact

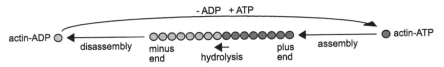

Fig. 1.2 An actin filament of ≈ 7 nm diameter is dynamically assembled from actin monomers. Actin-ATP molecules bind more strongly to each other. Upon ATP hydrolysis, actin-ADP molecules fall apart. Net dis-/assembly at the minus/plus end leads to a cyclic process called treadmilling

1.2.1   Actin Filaments Form Flexible Meshes

Actin filaments with a diameter of ≈ 7 nm are chains of actin monomers. Actin monomers exist in an ATP-bound state and an ADP-bound state. If ATP is bound to actin, actin-ATP molecules start to form filaments. When ATP is hydrolysed to ADP, the actin-ADP molecules fall apart from the filament, because the binding strength between actin-ATP molecules is stronger than between actin-ADP molecules. The exchange of ADP with ATP can only occur in free actin-ADP monomers. The dynamic equilibrium of actin filament assembly and disassembly depends on the availability of free actin monomers and the velocity of the ATP hydrolysis in actin-ATP molecules. At one end of the actin filament, ATP hydrolysis is faster than addition of monomers. This end shrinks and is called the minus end. At the plus end on the other side, the rate of newly added monomers is faster than ATP hydrolysis, which results in net growth of the filament in this direction. This polar structure of plus end and minus end emerging from these reactions is known as treadmilling (Fig. 1.2).

Fifty per cent of all actin molecules in a cell exist as monomers. To prevent these free molecules from spontaneous polymerization, they are bound to thymosin and profilin. Dynamic dis-/assembly of actin filaments is further guided by additional capping and cross-linking proteins that stabilize actin filaments. Actin filaments are an integral part of contractile bundles in the cytoplasm that connect different spots of the plasma membrane thereby giving the cell its characteristic shape. Dynamic de-/formation of actin filaments is required in filopodia, which are cellular feet to move forward. These aspects highlight the important structural and dynamic properties of actin filaments in cellular physiology.

1.2.2   Intermediate Filaments for Robustness

These filaments are called "intermediate" because, when they were first discovered in smooth muscles, the 10 nm diameter was between the diameter of thin actin filaments (see Sect. 1.2.1) and thick myosin filaments (see Sect. 1.2.4). The intermediate filaments form "rope teams", which are the toughest and most physically endurable cellular elements. They not only form a network throughout the cytoplasm but also support the nuclear envelope and cell-cell contacts called desmosomes (see Sect. 1.4). The structure of intermediate filaments reminds of a rope with many long strands that are twisted together. Each of these single ropes is an α-helical fibril with an N-terminal head region and a C-terminal tail region that spans a total length of 48 nm. Head and tail regions are heterogeneous and determine interactions with cytoplasmic proteins. Each of two ropes is twined together to form a pair. Two of these dimers are staggered to form a tetramer. Eight of these tetramers in turn can be packed together to form a non-covalently bound strand that assembles the final intermediate filament.

The main purpose of intermediate filaments is to make the cells more robust against mechanical stress such as squeezing or tearing. There are four major classes of intermediate filaments.

- Keratin filaments characterize the different thin tissue layers on the outer and inner surfaces of our body, called epithelia. The cells of each epithelium have an individual mixture of keratin filaments that span the cellular sheet horizontally through desmosomes (see Sect. 1.4.2).
- Vimentin and related filaments are mainly found in connective tissue, muscle cells and glial cells of the nervous system. Vimentin structures can be reinforced or cross-linked by accessory proteins such as plectin.
- Neurofilaments, as the name implies, are an integral part of nerve cells. Mutations that cause disruption of neurofilaments lead to neurodegenerative disorders.
- Nuclear lamina builds up the nuclear envelope. Tetramers of lamina can be quickly assembled and disassembled during cell cycle (see Chap. 7).

1.2.3   Microtubules as Tracks for Intracellular Transportation

Hollow tubes with a diameter of $\approx$ 25 nm are called microtubules. These tubes are built from subunits that exist as α-tubulin and β-tubulin forming heterodimers. These pairs of α- and β-tubulin exhibit a polarity with β-tubulin always pointing to the plus end and α-tubulin indicating the minus end in the opposite direction. The microtubule is dis-/assembled at the minus/plus end by dis-/assembly of tubulin. A chain of tubulin heterodimers forms a protofilament, and 13 protofilaments compose the wall of the tube surrounding the lumen (Fig. 1.3).

Microtubules form tracks from the cell's centre to the periphery for intracellular transport. The centre from which they grow out is called centrosome consisting of a pair of centrioles. This structure contains γ-tubulin that forms a ring thereby serving as a starting and anchoring point for the minus end of microtubules. The switching between the assembly and the disassembly of microtubules is known as dynamic instability (growing back and forth). Capping proteins prevent microtubules from

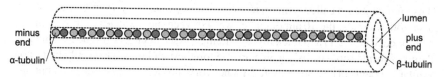

Fig. 1.3 A microtubule is a hollow tube of $\approx$ 25 nm diameter that is built from heterodimers of α- and β-tubulins. Long chains of these dimers form a protofilament, and 13 of these surround the lumen of a microtubule that exhibits a polar orientation from the minus to the plus end

shrinking at the plus end. Such selective stabilization can polarize a cell if capping proteins are provided at restricted areas, e.g. a particular site at the cell membrane. Microtubules can form cilia, 0.25 µm thin hair-like structures that sweep mucous (see Sect. 8.2) in our respiratory tract, and flagella, that propel sperm cells towards the egg. A ubiquitous function of microtubules is the formation of the mitotic spindle (see Chap. 7) and organization of the cell interior by allowing transport processes catalysed by motor proteins.

1.2.4   Motor Proteins to Position Cellular Cargo

One of the best-studied processes involving the cell's skeleton and motor proteins is muscle contraction involving actin filaments and myosin molecules. Myosin motor proteins exist as two subfamilies. Myosin-I is present in all cell types for vesicular transport (see Sect. 5.3). Myosin-II is only present in muscle cells. Each of the transport processes is actin-dependent and uses ATP as a source of energy. Most important for the cellular architecture are transport processes along microtubules.

The two most important families of motor proteins for the transport of cargo inside the cell are dyneins and kinesins. They both consist of two globular heads capable of tubulin binding and a tail domain specifically carrying the cargo. Motor proteins move forward by conformational changes. The energy for these movements is obtained by ATP hydrolysis. Dyneins move towards the minus end of microtubules at the cell's centre, whereas kinesins move towards the plus end of microtubules at the cell's periphery. The Golgi apparatus is positioned by dyneins around the nucleus, while the endoplasmatic reticulum (ER) is stretched throughout the cell by kinesins. The function of these cellular organs will be explained in the subsequent section.

1.3   Organelles of a Cell

Like our body, a eukaryotic cell contains organs, which are sites within the organism carrying out specific tasks. Cellular organs are called organelles and are introduced in the following.

- The **nucleus** is the central compartment of every eukaryotic cell because it stores DNA. The hereditary information is segregated from the cytoplasm by two nuclear membranes. Pores within the envelope allow the shuttling of molecules between the two cellular compartments. Most cells only contain a single nucleus. However, some cell types, such as liver cells, are binucleated and thus contain two nuclei.
- The **mitochondrion** is also surrounded by two membrane layers. The inner membrane is highly folded to increase the surface for the oxidation of molecules to produce energy in the form of adenosine triphosphate (ATP). In this process, oxygen is consumed and carbon dioxide is produced. A cell usually contains several mitochondria to fuel all cellular processes with ATP. Mitochondria are referred

to as cellular power plants, which make the cell breathe. Mitochondria contain their own DNA and divide to reproduce independently of the cell. Due to the two membranes, their size, function and separate genetic information, mitochondria are believed to be derived from prokaryotes, which were engulfed by other cells creating a symbiotic relationship. Symbiosis means that two organisms live together with mutual benefits.

- The **ribosome** is the protein synthesis machinery where messenger transcripts of the genetic code are translated into polypeptide chains to form functional proteins (see Sect. 3.3.2).
- A **chloroplast** is similarly organized as a mitochondrion. It is surrounded by two membranes, contains its own DNA, divides to reproduce and is derived from procaryotes. Chloroplasts are parts of plant cells and algae. They harness the energy from sunlight to build up sugar molecules. Therefore they contain stacks of membranes full of the green pigment chlorophyll. A side product of sugar manufacturing is oxygen. Sugars and oxygen can both be used then by mitochondria to produce energy.
- The **endoplasmatic reticulum** (ER) is a maze of interconnected compartments surrounded by a single membrane. The ER is the site where most membrane components and molecules destined for export from the cell are made.
- The **Golgi apparatus** (named after its discoverer Camillo Golgi) is an intricate network for sorting and barcoding of molecules for export from the cell or intracellular shipping (see Sect. 5.3.1).
- Cargo is transported from one cellular compartment to another in **vesicles**. In a special membrane-enclosed vesicle called **lysosome**, digestion of cellular components takes place. Breaking down molecules in a separate compartment is important for the release of nutrients, the recycling of building blocks or the excretion of unwanted substances. The **peroxisome** is another specialized membrane-enclosed vesicle to keep the cell from reactions involving hydrogen peroxide or other toxic substances. Some cell types are defined by their specific vesicles, e.g. neutrophils of the immune system.

A useful representation of these organelles can be found under https://smart.servier.com/.

1.4   Cell-To-Cell Contacts

Unicellular organisms are well equipped with their organelles to maintain life. In multi-cellular ensembles, cells have to interact and communicate with each other for vital functions of cellular layers. In tissues, cell-to-cell contacts serve as both, stabilizing elements and channels for information exchange. As they join cells, they are called "junctions". Three different classes of cellular junctions can be discriminated that will be introduced in the following.

1.4.1   Occluding Junctions as Diffusion Barriers

Occluding junctions, as the name implies, close up cellular layers. An epithelium is a sheet of cells that are all horizontally connected to separate outer medium from underlying cells. The border has to be leak-proof, e.g. for the fluids that lubricate the inner gut lining (see Sect. 8.1.1). The sealing is achieved by **tight junctions**, which adhere neighbouring cells so that aqueous solutions cannot diffuse between them. The two most important components of tight junctions are the transmembrane-proteins occludin and claudin. The function of tightly joined epithelial cells as a border implies a structural feature. Epithelial cells are polarized with their top side being different from their bottom side. In the gut, the apical (i.e. upper) side of the epithelial cells mainly contains the tight junctions plus microvilli, which form a cellular brush to increase the surface for uptake of nutrients. Other cells of this epithelium contain many vesicles on their apical side for the production of lubricious substances that flush the inner surface of the gut. The cytoskeleton of the basal (i.e. lower) side of these epithelial cells is connected to an extracellular matrix forming the basal lamina.

1.4.2   Anchoring Junctions for Adhesion

While tight junctions create a diffusion barrier at the apical side of the cells, **anchoring junctions** hold together the entire epithelium by mechanically attaching the cells with each other and with the basal lamina. Two classes of anchoring junctions can be discriminated: adherens junctions and desmosomes. Both rely on cadherin molecules, which can bind to one another. A cadherin molecule connects to the skeleton of one cell, spans the plasma membrane and connects extracellularly in the presence of Ca^{2+} ions to a cadherin molecule of a neighbouring cell. The name cadherin is derived from "calcium-dependent adhesion". Adherens junctions and desmosomes rely on different types of cadherin molecules but their main difference originates from the component of the cytoskeleton they are bound to. Adherens junctions are connected to bundles of actin filaments, whereas desmosomes are linked with keratin molecules. The actin filaments of adheren junctions connect all cells along the epithelium thereby forming an adhesion belt on the apical side capable of contractions. Keratin filaments in desmosomes tightly anchor cells to one another. If keratin is connected to the basal lamina, the junction is made by integrins not cadherins and the junction is called hemidesmosome.

1.4.3   Gap Junctions for Communication

Tight junctions and anchoring junctions keep the cells of an epithelium together, but **gap junctions** form tunnels that connect the cytoplasms of neighbouring cells by bridging the plasma membrane. So-called connexon channels of each six connexin subunits span the cellular membrane and align to connexon channels of adjacent cells. The pore through two connexon tubes allows direct exchange of soluble molecules between cells and thus couples them metabolically. The flux of electrically charged ions through gap junctions becomes important for instance in the transmission of neuronal signals in the retina [107].

Plant cells do only have one type of cell-to-cell contact, namely **plasmodesmata**, which penetrate the otherwise stabilizing cell walls and allow communication.

Summary

- Cells are mainly composed of carbohydrates, lipids, proteins and nucleic acids.
- The cellular skeleton contains three types of protein filaments: Actins, intermediate filaments and microtubules.
- The compartments of a cell are its organelles for specific reactions and functions.
- Cell-to-cell contacts allow both, tight closure of cellular layers and exchange between individual cells.

DNA & RNA & Associated Proteins

2

Contents

Introduction

The architecture of the cell needs a blueprint. Information on molecular components has to be encoded in the cell. It is written on long chains of nucleic acids. The discovery of deoxyribonucleic acid (DNA) as hereditary material, its structure and its maintenance represent central cornerstones for modern biology that paved the way for deciphering the human genome in 2001, and nowadays, high-throughput sequencing of genomes from bacteria, yeast, worms and fruit flies to cancer patients [41]. DNA sequence information has certain properties that allow being passed on from generation to generation. First, DNA is modular. Just a few simple building blocks make up long chains of information with sections called gene, from the Greek term *génos* meaning origin or offspring.

The long DNA chains are mostly wound up and condensed in chromosomes that define our sex and store all genetic information to be passed on to the next generation. The sequence of the hereditary material constitutes the genetic code, which is written as the language for the programme of life. But all the encoded information were useless if it could not be propagated. The molecular structure allows for the second important feature: DNA can be replicated. Building blocks are fused with a stretch of DNA complementary to a template strand thereby copy-pasting genetic information. If no errors occur within the process, the generated sister molecules are a duplicate of the original DNA.

The modularity and replication of DNA are translated to proteins that exert certain functions in cells. An intermediate product of protein production from the DNA blueprint is ribonucleic acid (RNA) molecules (DNA → RNA → protein). RNA

© Springer-Verlag GmbH Germany, part of Springer Nature 2022
L. Adlung, *Cell and Molecular Biology for Non-Biologists*,
https://doi.org/10.1007/978-3-662-65357-9_2

molecules share modular features of DNA. Besides being a DNA transcript for proteins, RNA has the ability to directly fulfil regulatory tasks. RNA molecules can either facilitate protein production as structural components of protein synthesis machinery or block this process by specific binding to RNA transcripts. These diverse features fuelled speculation that RNA could have been the central molecule for the origin of life [90]. However, the emergence of DNA and RNA is not mutually exclusive, because they share a common modular sequence structure (Fig. 2.1), which forms the virtual basis for all processes of modern biology.

2.1   Structural Elements

Most of the cell's biochemical components exhibit modularity because the assembly of complex structures from a small set of building blocks allows diversity of products and sustainable management of resources (see Sect. 1.1). The same holds true for DNA and RNA, which are made up from sugars that are linked to bases and connected by phosphate bridges (Fig. 2.1). The main difference between DNA and RNA arises from the type of sugar molecule that is used, which is implied by the name. For DNA, it is a Desoxyribose while for RNA, it is a Ribose. These sugars are pentoses because they contain five carbon atoms. DNA is more stable than RNA because the desoxyribose contains only a hydrogen atom at the 2' carbon, whereas the ribose exhibits a hydroxyl group that is much more prone to hydrolysis. DNA and RNA also differ with respect to their nitrogenous bases. There are five different nucleobases, and each one can be connected to the 1' carbon of the sugar residue in the macromolecule. Adenine, cytosine and guanine are part of both, DNA and RNA. Uracil is just incorporated into RNA, while thymine is solely part of DNA. Adenine and guanine belong to the **purine** bases. Cytosine, thymine and uracil belong to the **pyrimidine** bases. Sugar and base together are called **nucleoside**. If a phosphate group is added at the 5' carbon of the nucleoside's sugar, a **nucleotide** is formed. Depending on the sugar and the base that are connected, the nucleosides and respective nucleotides are called as given in Table 2.1.

The phosphate group that is added to the nucleoside for the formation of a nucleotide is derived from phosphoric acid, which loses its protons at physiological pH and is therefore negatively charged. The phosphate from phosphoric acid makes the nucleotide with its nucleobase a nucleic acid, too. The prefix "nucle-" originates from the fact that the material was first isolated from nuclei of white blood cells [108]. Nucleotides are the building blocks of DNA and RNA. Long stretches are created by phosphate-ester linkages. The phosphate group of a single nucleotide is

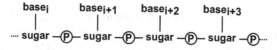

Fig. 2.1 Modular structure of DNA and RNA. Sugar molecules are linked via phosphate bridges and connected to different bases

Table 2.1 Composition, names and abbreviations for building blocks of DNA and RNA

Base	Nucleoside	Nucleotide (Abbreviation)
Purine		
Adenine	Adenosine	Adenosine 5'-monophosphate (5'-AMP)
	Deoxyadenosine	Deoxyadenosine 5'-monophosphate (5'-dAMP)
Guanine	Guanosine	Guanosine 5'-monophosphate (5'-GMP)
	Deoxyguanosine	Deoxyguanosine 5'-monophosphate (5'-dGMP)
Pyrimidine		
Cytosine	Cytidine	Cytidine 5'-monophosphate (5'-CMP)
	Deoxycytidine	Deoxycytidine 5'-monophosphate (5'-dCMP)
Thymine	Deoxythymidine	Deoxythymindine 5'-monophosphate (5'-dTMP)
Uracil	Uridine	Uridine 5'-monophosphate (5'-UMP)

always added to 5' carbon of the sugar and can form a phosphodiester with the 3' hydroxyl group of another nucleotide.

When the frequency of nucleosides in isolated DNA was analysed, adenosine (A) occurred always in similar amounts as thymidine (T) and cytosine (C) appeared as often as guanosine (G). These phenomena are meanwhile known as Chargaff's rules:

$$\frac{\#A}{\#T} = \frac{\#C}{\#G} = 1 \quad \Leftrightarrow \quad \#A = \#T \; ; \; \#C = \#G.$$

It was concluded that pairs of A:T, T:A, C:G and G:C make up the DNA. The purine base adenine can form two hydrogen bonds with the pyrimidine base thymine whereas three hydrogen bonds can be formed between the purine base guanine with the pyrimidine base cytosine. One strand of DNA is thus paired with another opposing strand. The sequences are referred to as **complementary** to each other, e. g.:

5' AGCCTGTAC 3'
3' TCGGACATG 5'

The sequence of nucleosides is linked by phosphodiesters creating a regular "back-bone" of the DNA molecule. The phosphate bridges are formed between the 3' hydroxyl group of a nucleotide and the 5' phosphate group of a subsequent nucleotide. If not indicated otherwise, sequences are usually given in the 5' → 3' direction. As the phosphodiester bonds at the backbone of the DNA molecule are negatively charged at physiological pH, they form a polar outer shell that protects the hydrophobic organic bases at the inner core of the molecule in aqueous solution of the nucleus. The 3D structure of the DNA molecule is a **double helix**, in which the two opposing DNA strands are held together firmly by all the hydrogen bonds between the base pairs. There are three different arrangements of DNA double helices: A-DNA, B-DNA and Z-DNA. These forms vary in their number of base pairs per circle, the size of the grooves and whether the spiral is left-handed or right-handed depending on the sequence modifications and the chemical properties of the solvent. The most common structure under physiological conditions inside cells is B-DNA. RNA exhibits much more diverse structures ranging from helices to loops and hairpins.

2.2   Organization of Genetic Information

2.2.1   Gene Elements, Histones and Chromatin

Given the size of the human genome with $\approx 3 \times 10^9$ base pairs (summing up to a length of more than a metre [88]), our DNA cannot just be loosely packed within the nucleus but needs to be tightly wrapped to fit in. DNA is coiled around a core element of $4 \times 2 = 8$ **histones**. A total of 147 base pairs of DNA fit 1.7 $\times$ around such an octamer of histone proteins. This constitutes a structure similar to "beads on a string" (Fig. 2.2). Together, DNA and histones constitute a nucleosome, which is a dynamically regulated, modular structure [71], and represents the basic unit of chromatin. The regulation of the nucleosomes involves histone modifications, e.g. acetylation or methylation of histone tails (Fig. 2.2). As these modifications are not directly readable from the genetic code, they are summarized as epigenetic information (see Sect. 4.4). The histone modifications therefore represent a regulatory layer on top of the DNA sequence. They determine the structural arrangement of DNA and proteins, which is summarized as **chromatin.**

Depending on the cellular state, chromatin can exist in different conformations. If the chromatin is accessible, it is referred to as "open chromatin". This means that stretches of DNA are not (tightly) bound to histones and can thus be read, replicated and transcribed. On the contrary, in regions of "closed chromatin", nucleosomes are tightly packed and cannot be accessed unless they are modified and rearranged. The efficient wrapping of DNA around histones within closed chromatin reduces the volume that the macromolecules overall need. This arrangement gives rise to a particular structure called "heterochromatin" as opposed to "euchromatin" with mostly open chromatin.

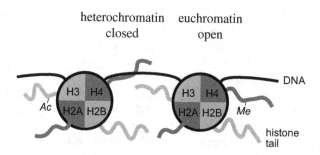

Fig. 2.2 Histones are proteins with each two of the four subunits: H2A, H2B, H3 and H4. DNA is wrapped around. Histone tails can be chemically modified, e.g. acetylation (*Ac*) or methylation (*Me*)

The heterochromatin revealed a higher level of a structural organization unit of the genetic information: the chromosome.

2.2.2   Chromosomes, Genes and Heredity

The word chromosome is derived from Greek and means "coloured body" because chromosomes were studied during cell division (see Sect. 7.3.2) by staining with certain dyes. The division is a part of the cell's process of duplication (i.e. replication), which is an essential step in life as such. In this process, two states of the chromosome can be discriminated by their 3D conformation. The **metaphase chromosome** is the state during the cell division that was discovered first as it is very condensed and thus well structured (Fig. 2.3). The geometry is not well understood, but subunits can be described. The metaphase chromosome is composed of two bar-shaped structures, called **chromatids**. Each of the two chromatids of a chromosome contains a copy of the same DNA. The pair of chromatids of a chromosome is referred to as sister chromatids. They are held together and linked by the centromere located in the centre of the metaphase chromosome, dividing each chromatid into a short arm and a long arm. The terminal ends of each chromatid are called telomeres, which are important to prevent shortening of chromosomes during replication (see Sect. 2.3), which takes place prior to the metaphase of the cell division process.

Most of the time, the chromosome exists in the interphase state. The interphase is the time between two cell divisions, as derived from the Latin term *inter* meaning "between". The **interphase chromosome** is characterized by a loose arrangement of DNA, less condensed than metaphase chromosomes. Despite being more loosely arranged, the DNA of interphase chromosomes is not randomly distributed in the nucleus. Instead, there are chromosome territories in which regions of the DNA are predominantly located during the interphase [28]. These spatially restricted areas can be visualized by means of fluorescence *in situ* hybridization.

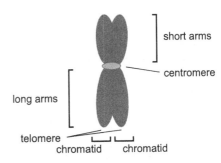

Fig. 2.3 Schematic conformation of a metaphase chromosome. Two chromatids with identical DNA constitute the condensed structure that is held together by the centromere that divides the chromatids into a short arm and a long arm. Terminal ends are called telomeres

Fluorescence *in situ* Hybridization aka. FISH

DNA or RNA is labelled in its original location (i.e. "*in situ*") by binding to a fluorescent probe that can be detected under a fluorescence microscope. Cells are fixed and short nucleotide sequences are injected as labels to bind through hybridization (i.e. the formation of hydrogen bonds) to complementary regions of the cellular nucleotides in their original regions. The labels are not mere nucleotide stretches, but they are most commonly coupled to fluorescent dyes, which emit light when excited by a laser at a certain wavelength, which can be detected under a fluorescence microscope to visualize the localization of DNA or RNA inside a cell as for the diagnosis of breast cancer [88].

Another method to investigate distinct sites of a chromosome is called chromosome conformation capture ("3C"). The idea behind this technology is that distant regions of a chromosome are brought into spatial proximity and interact if they are active as open (eu)chromatin at the same time. Then, these regions can be stuck together, in a process called ligation (derived from Latin *ligare*, which means to tie). These stretches of DNA being regulated together can be between 200 kilobases (kb) and 1.5 megabases (Mb) long. They are referred to as **topological domains**. Two topological domains are separated by so-called boundary elements [29]. The precise regulation of chromosomes' structure is a precondition for the preservation of a cell's hereditary information. Every human cell contains two copies of each chromosome. Ploidy refers to the number of complete sets of chromosomes, and since human cells have *two* copies, they are referred to as *di*ploid. In total, there are 46 chromosomes, consisting of a pair of sex chromosomes, which determines the human sex (XX = female, XY = male; even though there are several intermediate states), and additional 22 pairs of body chromosomes called autosomes.

The most important functional subunit of hereditary material is a **gene**. A gene is a stretch of DNA, that encodes the information to build a molecule with a particular function. Molecules encoded by genes can be ensembles of either nucleotides of an RNA or amino acids of a protein. Every gene is located at a certain position on a particular chromosome. The gene's position (i.e. location) is called its **locus**. The gene exists on both duplicates of the pair of chromosome in that locus. The gene can be present in different versions, which are called **alleles**. For instance, the *ABO* gene is located on the human chromosome number 9. The *ABO* gene has three different alleles that encode our blood types: A, B and 0. If one copy of the pair of chromosome 9 harbours the A allele, the other copy of the pair of chromosome 9 harbours the B allele, the blood type is AB. In this case, the two alleles of the *ABO* gene are different from one another, so they are called **heterozygous**. If the blood type is AA, BB or 00, the *ABO* gene is **homozygous** because the two copies of the gene exist in the same version.

The entire genetic composition of a cell or an organism is called their **genotype**. The physical outcome of this genetic composition (e.g. the blood type) is called the **phenotype**, which is a characteristic that can be passed on from one generation to another, and as such represents a heritable trait. Sometimes, one allele of a gene is dominant over the other one, so a single copy of this allele is sufficient to confer a certain phenotype. Otherwise, recessive alleles, which are inferior to the dominant forms, are required as two copies to show the phenotype. In our blood type example, the A and B alleles are dominant over the 0 allele. Consequently, A0 and 0A result in blood type A, B0 and 0B result in blood type B and only 00 results in blood type 0.

2.3   The Genetic Code

A set of rules is required for the conversion of genetic information into phenotypic output. The link between the nucleotide sequence of the gene and the amino acid sequence of the protein is encrypted by what is called the **genetic code**. Genetic information is storable within cells and can be inherited from one generation to another. To transfer the information stored in the DNA sequence to functional proteins an intermediate layer is involved: RNA. The conversion from the DNA to the RNA sequence is called transcription. It is straightforward as both macromolecules are closely related and structurally very similar. Sequences are complementary to each other and use the same letters A, G and C, while the DNA nucleotide thymidine (T) is replaced by uridine (U) in RNA.

The conversion from a nucleotide sequence to an amino acid sequence is rather complex because there are only four different nucleotides $\in \{A; U; G; C\}$, which need to be mapped to 20 different amino acids to form proteins. Information encoded by RNA nucleotides are processed as triplets. Given the four different nucleotides, there are $4 \times 4 \times 4 = 64$ possible triplets. Since all of them are being used, the information is redundant, because 64 triplets encode only 20 different amino acids. The genetic code is thus degenerated. The reason for this extent of complexity and redundancy is three-fold [102]:

- Diversity of amino acids required.
- Robustness against errors during information processing.
- Minimization of metabolic costs.

The triplets are called **codons** and, of course, 20 different codons are required at least to encode for the 20 amino acids, which serve as building blocks for proteins (see Sect. 1.1). Chemically similar amino acids are encoded by triplets that are confused by chance. The readout of the codons is an erroneous process and if a nucleotide is likely to be mistaken, the corresponding correct and incorrect amino acids are likely to be the same or at least comparable in their polarity, side-chain characteristics or other biophysical properties. Too many redundant codons do not exist because their maintenance is metabolically costly [98]. E.g. methionine and tryptophan are encoded by only a single codon. These three aspects explained above steered the evolution of the genetic code and rendered it modular (Table 2.2). With only a few exceptions, for instance, in *Mycoplasma*, the genetic code is universal.

Table 2.2 Amino acids are encoded by triplets of nucleotides (in bold). Amino acids and codons are given by the following one-letter code: F=phenylalanine, L=leucine, I=isoleucine, M=methionine and start codon, V=valine, S=serine, P=proline, T=threonine, A=alanine, Y=tyrosine, Z= stop codon, H=histidine, G=glutamine, A=asparagine, K=lysine, D=aspartic acid, E=glutamic acid, C=cyteine, W=tryptophan, R=arginine, G=glycine

1st	U				C				A				G			
2nd																
U	F	F	L	L	L	L	L	L	I	I	I	M	V	V	V	V
C	S	S	S	S	P	P	P	P	T	T	T	T	A	A	A	A
A	Y	Y	Z	Z	H	H	Q	Q	N	N	K	K	D	D	E	E
G	C	C	Z	W	R	R	R	R	S	S	R	R	G	G	G	G
3rd	U	C	A	G	U	C	A	G	U	C	A	G	U	C	A	G

2.4 DNA Replication

For maintenance of DNA integrity and genomic stability throughout generations, DNA has to be replicated robustly. The production of daughter cells requires DNA synthesis in the so-called S(ynthesis) phase of the cell cycle (see Chap. 7) prior to cell division. DNA replication is precisely regulated in space and time [35]. Starting points for these processes within the genome are the origins of replication. The human genome contains roughly 10,000 of these origins, and thousands can be activated simultaneously within a single cell but only once within a single round of DNA replication. Otherwise certain regions of the genome would be amplified causing genomic instability.

An **origin of replication (ORI)** is defined as a site within the genome where DNA replication starts. Every origin consists of two distinct elements. One region is recognized by a molecular machinery to form the pre-replication complex. Another region recruits the enzyme for DNA synthesis, namely DNA polymerase I in eukaryotes. Due to these two distinct sites at the origin, the DNA replication process can be discretized into two subsequent events:

1. **Licensing**: The origin of replication is recognized by an origin recognition complex that recruits additional proteins for the formation of the pre-replication complex.
2. **Firing**: The enzymes required for DNA replication are directed to the sites for DNA synthesis in a cell-cycle-dependent manner.

The DNA double helix has to be unwound for replication. ORIs are AT-rich because only two hydrogen bonds exist between adenosines and thymidines and therefore these nucleotides can be easily prised apart. The double strand needs to be opened like a zipper thereby forming a **replication fork** so that the DNA sequence can be read and copied. To unzip DNA, hydrogen bonds between the two DNA strands of the double helix have to be removed (Fig. 2.4). As single-stranded DNA is less stable, these unzipped strands have to be stabilized to prevent their degradation or

Fig. 2.4 DNA double helix needs to be unzipped for replication

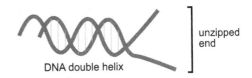

unzipped end

DNA double helix

reformation of the double strand. During the copying of the DNA template, molecules precede the replication machinery, which clear the way and adhere the complex to the strand. Thus, many specific factors are involved in DNA replication.

2.4.1   Molecular Components Involved in DNA Replication

- **Helicase** unwinds double-stranded DNA under ATP hydrolysis.
- **Topoisomerase** relieves tension of twisted DNA.
- **Single-strand binding protein (SSB)** stabilizes single strands of unwound DNA.
- **Primase** provides RNA primer as a starting point for DNA synthesis
- **DNA polymerase** synthesizes DNA as a complementary copy of DNA template strand.

DNA replication begins with a helicase that prises the two strands of the DNA double helix apart. Energy from ATP hydrolysis is required for this unzipping enzyme to be propelled forward. While the helicase opens the replication fork, the twisting of the yet unwound DNA double helix in front of the helicase creates a tension that needs to be relieved for the replication machinery to proceed. DNA is cut partially to prevent it from being wound up more tightly. The enzymes responsible for creating nicks and resealing upon unwinding are topoisomerases. Once the DNA replication fork is opened up, single-strand binding proteins ensure that the fork is not prematurely collapsing. For the actual DNA synthesis to begin, it needs some starting material, which is an RNA primer provided by a primase. This enzyme belongs to the family of RNA polymerases because it synthesizes RNA complementary to the DNA template. The DNA polymerase can elongate the RNA primer in the 3' direction with DNA complementary to the template strand. The DNA polymerase adds nucleoside 5'-triphosphates upon hydrolysis of pyrophosphate thereby creating a phosphodiester bond between the 5' phosphate group of the nucleotide and the free 3' hydroxyl group of the strand that is to be elongated.

The human DNA polymerase I processes ≈ 100 nucleotides per second. The nucleotides complementary to the template are energetically favoured and therefore bind faster. If there is an error, such as A not matched to T or C not matched to G, the DNA polymerase slows down and uses its **proof-reading activity**. The enzyme tries to immediately correct the wrong nucleotide by removing it and waiting for the binding of the nucleotide that is complementary to the template strand. If there is a correct match, the phosphodiester bond between the adjacent nucleotides is formed and the DNA polymerase proceeds in the 5'$\rightarrow$3' direction of the nascent DNA strand. Despite its proof-reading activity, the human DNA polymerase I creates an error every 10^7 nucleotides. This error rate adds up to ≈ 300 mismatches within

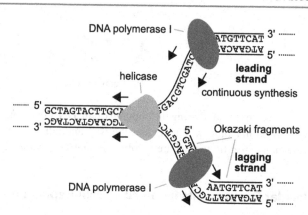

Fig. 2.5 For DNA replication, the double strand is unwound by the helicase creating a replication fork. On the leading strand, DNA polymerase I continuously synthesises DNA in 5'→3', whereas, on the lagging strand, Okazaki fragments are discontinuously added

a single round of replication of the cellular genome. To prevent these mismatches from being propagated throughout subsequent generations, additional DNA repair mechanisms exist (see Sect. 7.2.2). A so-called repair polymerase is also involved in filling the gaps that were created upon removal of the RNA primer sequence. When a nuclease cuts the RNA nucleotides out of the newly synthesized strand, the repair polymerase fuses DNA nucleotides in. A ligase links the 3' hydroxyl group of the synthesized DNA strand with the 5' phosphate of the inserted nucleotides via a phosphodiester bond, which seals the loose ends.

To replicate the entire genome of a cell, multiple DNA polymerase I molecules have to be present at every single replication fork. The sequences of the two strands of the DNA double helix are complementary to each other and their direction is anti-parallel. DNA synthesis by DNA polymerase I can only occur in 5'→3' direction of the nascent DNA strand. This implies that the replication machinery moves in opposite directions along the two loose ends of the replication fork. On one strand, DNA polymerase I follows the helicase towards the DNA region that is being unwound and DNA synthesis is continuous. On the other strand, DNA polymerase always approaches the end of the fork and its DNA synthesis can only occur discontinuously (Fig. 2.5). The fragments that are synthesized between the provided RNA primer and the end of the strand are named after their discoverer: **Okazaki fragments**. As the continuous DNA synthesis is faster than the discontinuous addition of Okazaki fragments, the strand of continuous DNA synthesis is considered the **leading strand**, while the other one is the **lagging strand**.

The problem at the very end of the chromosomes is that the RNA primer on the lagging strand cannot be replaced. To prevent chromosome shortening due to loss of this single-strand overhang, **telomerases** create a repetitive DNA sequence at the terminal ends of the chromosomes called **telomeres**. The repetitive DNA sequences rely on an RNA template that is provided by the telomerase itself. Attrition of telomerases declines telomere maintenance and is linked to ageing and disease [14]. Telomeres prevent the terminal sites of chromosomes from being recognized as double-strand breaks and unwanted recombination.

2.4.2 DNA Recombination to Secure Integrity and Diversity

Errors are made while the DNA is replicated. Ideally, an exact copy of the cellular DNA template is generated, but changes in the normal DNA sequence can occur. A deviation of the DNA sequence from the original sequence is called a **mutation**. An error in this case means, for instance, that an G(uanosine) is introduced opposite of an A(denosine), whereas T(hymidine) would be the correct complementary base. The observed frequency of such errors during DNA replication is around 10^{-4}. The observed frequency of observed mutations, however, is only 10^{-9} per copied nucleotide. This high accuracy stems from the "**proof-reading**" activity of the DNA polymerase, the enzyme that synthesizes new DNA. The proof-reading mechanism was studied not only in human cells but also in bacteria, mostly in the rod-shaped bacterium *Escherichia coli* (abbreviated *E. coli*). *E. coli* is called a "model organism" because it served traditionally as the main subject of experiments in molecular biology, and many experimental techniques were established with this bacterium. *E. coli* is a prokaryotic cell and has three different types of DNA polymerases, all of which with proof-reading activity. Two of the DNA polymerases in eukaryotic cells (e.g. human) are able to perform proof-reading.

The synthesis of DNA through both, prokaryotic and eukaryotic DNA polymerases, happens in 5' → 3' direction. Proof-reading is ensured by a 3' → 5' exonuclease activity, which means that the DNA polymerase can go back in the opposite direction of the newly synthesized strand and cut out the previously inserted, wrong nucleotides. Upon pause and removal of the incorrectly inserted nucleotide, the template strand of the DNA is copied again. If base pairing is happening correctly, the synthesis proceeds in 5' → 3' direction.

Besides copying errors, DNA is exposed to other damage-causing agents such as ultra-violet (UV) light that may ultimately lead to the generation of cancer (see Sect. 9.1.2). Mutations that affect only a single nucleotide, which is changed compared to the original (i.e. normal) DNA sequence, are referred to as point mutations. These point mutations will be propagated if they are not corrected. In addition to proof-reading, there are several DNA **excision-repair systems** to prevent mutations from being established in the genetic code. These systems all work in a similar way. The damaged segment of a single DNA strand is cut out, and the created gap is refilled by the DNA polymerase complementary to the original template. Three types of excision-repair systems are distinguished.

- Base excision-repair: Most common in humans is a point mutation changing the original C(ytosine) to T(hymidine) thereby creating an erroneous $\frac{G}{T}$ pair instead of the correct $\frac{G}{C}$ nucleotide pair. The incorrect pair is detected, because G and T do not fit as well together as G and C. It is hard for the DNA repair machinery to recognize whether the mutated nucleotide is the G or the T. But since the mutation from C to T is much more likely, the system evolved to replacing the T of the incorrect $\frac{G}{T}$ pair with a C by default.
- Mismatch excision repair: A general case of the base excision-repair that eliminates other point mutations, or insertions or deletions of several nucleotides.

Insertions mean pasting of more nucleotides in synthesized strand than the template. All these errors occur during DNA replication by DNA polymerase. It is not fully understood how the system discriminates the correct template from the erroneously synthesized strand. Some proteins detect and bind to the ill-shaped mismatch of incorrect nucleotide pairs, and subsequently guide the cutting and replacement.

- Nucleotide excision repair: This applies to single nucleotides that are chemically modified and then interact with neighbouring nucleotides of the same DNA strand. The prime example is a dimer (i.e. a pair) of two adjacent T(hymine)s in the same DNA strand. Since one T is chemically added to the subsequent T, it is also referred to as a "chemical adduct". The T + T structure results in a local distortion of the DNA double helix. A helicase is recruited to open and unwind this part of the helix, which facilitates the excision (i.e. cutting out) and replacement of the wrong T + T dimer.

Two other systems use DNA recombination instead of excision and repair to keep DNA integrity. DNA recombination is the exchange of DNA snippets between very similar regions, e.g. the same locus of the two copies of the same chromosome in a diploid cell, or following replication, (see above, Sect. 2.2.2). Recombination-based DNA repair happens when both strands of the DNA are ruptured as if one cuts a telephone wire (i.e. the DNA double helix) with a pair of scissors (usually caused by irradiation or drugs). The resulting disjoint structure is called a "DNA double-strand break". The loose ends of the DNA can cause gross structural rearrangements of the chromosome, which has to be prevented to keep the genetic information intact. DNA double-strand breaks can be repaired by two different mechanisms.

1. **Homologous recombination**: Overhangs are created by cutting out opposing strands from the two broken ends of the DNA double helix through the exonuclease activity of enzymes. The thus-created overhangs are then hybridizing (i.e. binding) to the homologous (i.e. similar) region of the other copy of the same chromosome. The broken ends are extended through DNA replication of the template strand from the other copy of the (homologous) chromosome. Once the former loose ends are connected, the junctions between the homologous chromosomes are cut (Fig. 2.6). This process leads to the repair of DNA double-strand breaks and involves exchange (i.e. recombination) of genetic information between the homologous chromosomes.
2. **Non-homologous end joining**: The loose ends of the broken DNA double strand are merged together with each other or with similar regions in the proximity of the same chromosome. This involves cropping of the terminal overhangs to create blunt ends, which leads to the loss of several nucleotide base pairs.

Once DNA integrity is ensured for replication of genetic information, the hereditary material can be decoded for the propagation of functional properties of the cells.

Fig. 2.6 A DNA
double-strand break (1) is cut
to generate overhangs (2),
which bind to the
complementary regions of
the other copy, and DNA
replication (dashed line)
completes the second strand
(3). Subsequent cutting of
X-like "holiday" structure
(4), and connecting the black
strands (5). 5' → 3'
direction indicated by
arrowheads

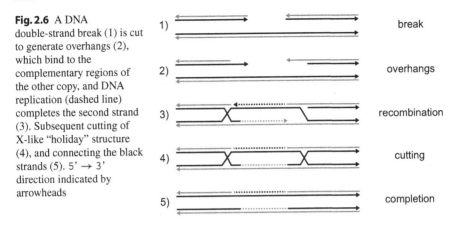

1) break
2) overhangs
3) recombination
4) cutting
5) completion

Summary

- Nucleotides are building blocks of DNA and RNA.
- Genetic information is encoded in DNA, which is wrapped around histone proteins to form chromatin structure in chromosomes.
- The genetic code degenerated because 64 possible combinations of three nucleotides encode for only 20 amino acids.
- DNA replication ensures integrity and maintenance of hereditary material through different proof-reading and repair mechanisms.

Transcription and Translation

3

Contents

Introduction

The mere maintenance of genetic information and its structure is insufficient to carry out cellular functions. Genetic information has to be *expressed*. Therefore, the information stored in the DNA sequence needs to be decoded, i.e. transcribed, into a messenger molecule consisting of RNA. Mostly, this message in turn needs to be translated into a molecule, i.e. a polypeptide, which, upon folding into a protein, exerts the function of the gene with a certain property (Fig. 3.1). Reading, writing and translating genetic information are fundamental cellular processes, and their main principles are evolutionary conserved. The molecular processes of gene *expression* can be investigated in bacteria as well as in human cells. Importantly, it is not sufficient to just replicate DNA, transcribe it to messenger RNA (mRNA) and translate this to protein. In biology, structure and function are intimately related. The protein at first is a steady chain, which has to fold into a three-dimensional (3D) structure. The protein cannot function properly unless its correct conformation is reached. Erroneously folded proteins can aggregate and plug the cell or do harm otherwise. That is, the reason why unfolded (i.e. incorrectly folded) proteins will be erased (i.e. degraded) from the cell.

With modern technologies it is possible to measure all cellular DNA, or RNA, or all proteins of a bulk of cells. If these entities are measured at this global scale, the processes are referred to as "-omics". For instance, the measurement of all genetic information in the form of DNA is referred to as genomics. The measurement of

© Springer-Verlag GmbH Germany, part of Springer Nature 2022
L. Adlung, *Cell and Molecular Biology for Non-Biologists*,
https://doi.org/10.1007/978-3-662-65357-9_3

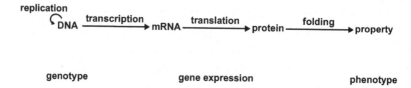

Fig. 3.1 The genetic information stored in DNA is replicated and transcribed into a messenger molecule of RNA (i.e. mRNA), which in turn is translated to a protein that needs to fold into a particular 3D conformation to obtain a structural and thus functional property. The process of gene expression links a genotype to a phenotype

all transcripts (i. e. mRNA) is referred to as transcriptomics. Measuring all proteins of cell populations is called proteomics. Analogously, the whole DNA of a cell is its genome, all transcripts are called the transcriptome, and all proteins make up the proteome. The sequence of DNA and RNA can be read by machines, thus the measurements of genome and transcriptome are referred to as "sequencing", whereas individual proteins are analysed based on their mass and electrical charge as ions. The proper interpretation of such datasets is still challenging, but it allows a global perspective on fundamental cellular processes of life.

3.1   The Genome is Read and Transcribed

To read and transcribe the genetic information, a reading device is needed. A molecular machine capable of reading and copying DNA is the DNA polymerase (see Sect. 2.4.1). A polymerase is an enzyme capable of synthesizing and outputting a long molecule of many units (i.e. a polymer, *poly* from Greek meaning "many"). In case of the DNA polymerase, the output is a DNA polymer because the DNA is copied. To exert the function of a gene, the genetic information in the form of DNA needs to be transcribed into messenger RNA transcripts first. The enzyme in this case has to output RNA, and is thus an **RNA polymerase**. Bacteria have only a single type of RNA polymerase, which, however, consists of several subunits with specific functions such as providing the primer (i.e. the first building block) for the synthesis of the RNA polymer.

The genome of eukaryotes is more complex, and so are the structures of their RNA polymerases. Humans, for instance, have at least three diverse types of RNA polymerases: I, II and III. The **messenger RNA (mRNA)** transcripts are generated among other RNA molecules (see Sect. 4.2) by RNA polymerase II. The entire process of transcribing DNA into mRNA and their translation into a protein, which folds into a functional structure, is referred to as **gene expression**, because the information encoded in a gene is thereby decoded and finally expressed.

3.1.1 Start the Transcription

The first step of gene expression is the initiation of the transcription of the DNA. A starting point has to be found to mount the process. The DNA sequence is indexed relative to the transcription start site (TSS). The TSS is the origin (0) and the DNA template strand to be transcribed is displayed in 3' → 5' direction. The left side of the TSS is referred to as "upstream" with a negative index number of nucleotide bases, and right to the TSS is the downstream sequence with positive index numbers of nucleotide bases. The region where transcription and thus gene expression is initiated is called **promoter** region (Fig. 3.2).

In bacteria, the module that detects the TSS is the σ factor, a protein tightly associated to the bacterial RNA polymerase. It binds to a **consensus** sequence upstream of the TSS to find the position to initiate transcription. A consensus is a sequence region that is very similar among different species and thus evolutionarily conserved. The consensus sequence for the σ factor to bind the DNA to be transcribed is around the positions −35 and −10 (Fig. 3.2). The spacing between these regions varies around 17 base pairs. The consensus sequence at −35 is TTGACA, and the consensus sequence at −10 is TATAAT. This does not mean that upstream of every TSS in bacteria this exact sequence can be found, but it is the most frequently observed sequence among bacterial species. For eukaryotes such as humans, the DNA-binding region for the initiation of transcription is around the position −25 upstream of the TSS. This sequence is less conserved than in bacteria, and it is referred to as the TATA box rather than giving a consensus sequence.

While in bacteria, the σ factor together with RNA polymerase is sufficient to initiate transcription, eukaryotes require additional proteins capable of DNA binding to recruit the RNA polymerase II; so-called **transcription factors**. The events happen in the following order:

1. The TATA box-binding protein (TBP) as subunit of the transcription factor TFIID (standing for *T*ranscription *F*actor, RNA polymerase *II*, *D*) recognizes the TATA box in the promoter region of the gene to be transcribed.
2. Other general transcription factors (TFIIB, TFIIE and TFIIH) are recruited to the DNA of the TATA box as well as the RNA polymerase II. Altogether, they form the so-called transcription initiation complex.
3. The transcription factor TFIIH serves as a helicase that unwinds the DNA powered by ATP hydrolysis. In addition, TFIIH chemically modifies the C-terminal domain (CTD), a side arm of RNA polymerase II.

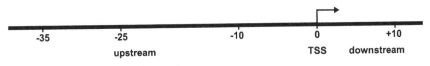

Fig. 3.2 Promoter region usually symbolized by a bent arrow around the transcription start site (TSS). DNA nucleotide positions in the direction of transcription are downstream and positively indexed, whereas the opposite, upstream, is negatively indexed

4. The RNA polymerase II synthesizes an RNA primer and changes its conformation to cause the release of the general transcription factors. Thereupon, transcription starts from the nucleotide at the position +1 of the DNA template strand.

3.1.2 Quality Control During Transcription

Besides its role in transcription initiation, the transcription factor TFIIH is also implied in transcription-coupled nucleotide excision repair (see Sect. 7.2.2). When RNA polymerase II becomes stalled at $T+T$ dimers, TFIIH is recruited again to unwind the DNA for repair. Once the transcription is successfully initiated, the RNA polymerase runs along the RNA polymerase runs along the DNA template strand in 3' → 5' direction and synthesizes a nascent polymer from complementary RNA nucleotides in 5' → 3' direction. This process is called **elongation**. In contrast to DNA polymerases, RNA polymerases lack proof-reading ability, because they do not have any 3' → 5' exonuclease activity. Instead, there is a kinetic proof-reading phenomenon, which applies during the elongation of mRNA synthesis. If an RNA nucleotide is not the correct one (i.e. not-complementary to the DNA template strand,) then the RNA polymerase slows down because hydrogen bonds cannot form properly between the nucleotides and no energy is released that fuels the process. The longer waiting time at this position increases the probability that the RNA nucleotide falls off again since the binding and the unbinding are in equilibrium. By chance, the wrong RNA nucleotide is released and replaced by the correct one, which causes the formation of hydrogen bonds, the release of energy and the subsequent acceleration of the elongation of the nascent RNA polymer. Due to the kinetic "proof-reading", the error rate of RNA polymerases is as low as $\approx 10^{-4}$ per nucleotide. This is much higher than the error rate of DNA polymerases, but in contrast to copied DNA, mRNA transcripts will not be inherited. Wrong mRNA molecules can cause only little damage to a cell. Being single-stranded, mRNA molecules will be rapidly hydrolysed (i.e. degraded). If the mRNA is translated to a wrong protein, there are additional steps of quality control for it to be removed (see Sect. 3.3.3) Despite eventual deceleration by the kinetic proof-reading, elongation continues until it is signalled to terminate.

3.2 Processing of Transcripts

3.2.1 Stop the Transcription

In prokaryotes, there are two different ways to terminate the transcription from DNA into mRNA by the RNA polymerase.

1. Rho-independent termination: The nascent RNA polymer forms secondary structures, such as a hairpin, through binding of repetitive regions that are complementary to each other within the same mRNA molecule. The secondary structure pulls the transcript away from the RNA polymerase (Fig. 3.3).

Fig. 3.3 An mRNA hairpin is formed between repetitive, complementary regions, that physically pull off the nascent mRNA molecule from the RNA polymerase and the DNA template strand, which is transcribed along its 3' → 5' direction

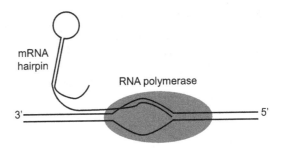

2. Rho-dependent termination: The RNA-binding protein Rho is recruited to the nascent RNA polymer and slides towards the RNA polymerase to unzip the hybrid between DNA template and mRNA transcript. Since hydrogen bonds between the complementary nucleotides have to be broken, energy is required, which is obtained through ATP hydrolysis by Rho. As such, Rho is a helicase that unwinds the DNA-mRNA helix, and an ATPase.

In eukaryotes, the nascent mRNA transcript is processed during its synthesis. Termination of transcription is regulated within this process.

3.2.2 Splice the Transcript

The processing of the eukaryotic mRNA involves three distinct steps, all of which start, while the DNA is still being transcribed in the nucleus (Fig. 3.4). The structure of a gene is important to understand the necessity of the processing. A gene sequence consists of coding and non-coding regions. Coding regions contain the blueprint for the protein that will be translated from the mRNA transcript. Non-coding regions have regulatory functions (see Sect. 4.2), but they do not contain information for the protein to be synthesized. Coding regions are called "**exons**", while the non-coding regions of a gene are so-called "**introns**". Introns are still part of the nascent mRNA strand synthesized from the DNA template by the RNA polymerase II, and have to be cut out during mRNA processing. Cutting out introns from the pre-(mature) mRNA, leaving a mature mRNA containing only exons is referred to as **splicing**. The prediction of introns and splicing sites from the mRNA sequence is a non-trivial task. There is a consensus sequence, which, however, can occur just by chance on the transcript without determining the splicing sites of the intron:

GU	A	YYYYYYYYY	AG
5' splicing site	branch point	poly-pyrimidine tract	3' splicing site

To cut out the intron, the adenosine at the branch point loops forward and binds to the 5' splicing site. The 3' end of the adjacent exon (to the left of the consensus sequence above) is subsequently free and can connect to the exon at the opposing end of the intron. The joining of the two exons releases the intermediate intron as a lariat structure with a loop between the 5' splicing site and the branch point, and the poly-

Fig. 3.4 RNA processing
involves the addition of the
5' cap, splicing of introns so
that only exons remain in the
mRNA transcript, and the
addition of the poly-A tail

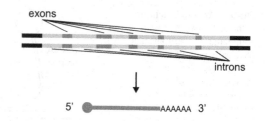

pyrimidine tract as a tail. The exons are finally connected in a continuous sequence, and the excised intron is recycled into single nucleotides for another mRNA polymer.

In addition to splicing, the nascent mRNA polymer receives a 5' cap and a poly-A(denosine) tail at its 3' end. Once the synthesis of the 3' end of the mRNA is completed, the nascent polymer is cleaved from the RNA polymerase II. Subsequently, a poly-A polymerase is recruited to the mRNA to catalyse the synthesis of a poly-A tail at the 3' end of the mRNA. Once the poly-A tail is generated, it is bound by poly-A binding proteins that stabilize the mRNA. The 5' cap is a 7-methyl G(uanosine), which is linked through a 5'–5' triphosphate bridge to the first nucleotide at the 5' end of the mRNA. The 5' cap stabilizes the single-stranded RNA molecule, because it protects it from 5' → 3' exonucleases, which degrade incorrectly processed mRNA molecules. In addition, the 5' cap guides the mRNA orientation, facilitates its export from the nucleus to the cytoplasm and initiates the translation of the mRNA molecule (see Sect. 3.3.1).

3.3   Transcripts are Translated into Polypeptides

The structure of proteins is important to understand the translation of mRNA transcripts to proteins. The basic, modular unit of a protein is an amino acid. An amino acid has a conserved core structure with an amino (NH_2) group, a carboxyl(ic acid, i.e. COOH) group, and a hydrogen (H) group around the central carbon atom (C) as well as a variable residue ("R") that determines the type of the amino acid. The amino group of one amino acid can bind to the carboxyl group of another amino acid, thereby forming a **peptide** bond. A molecule consisting of *two* amino acids that are linked by a peptide bond, is called a *di*peptide. Depending on the number of amino acids in such a molecular chain, the molecule is called a di- (2), tri- (3) or poly- (>10) peptide (Fig. 3.5). Every peptide has an N-terminus (i.e. the end with the unbound NH_2 group), and a C-terminus (i.e. the end with the unbound COOH group). Polypeptide chains that are ultimately folded into a 3D structure are mature proteins. To translate the mRNA transcript into the encoded amino acid sequence, another molecular assembly line (similar to the transcription machinery) has to be initiated.

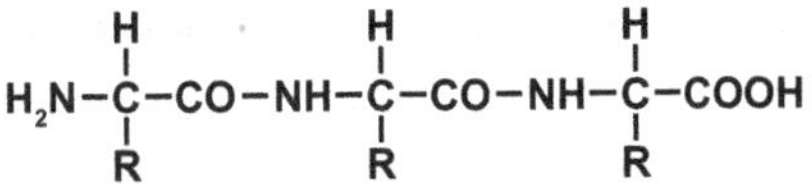

Fig. 3.5 A tripeptide consisting of three amino acids characterized by their variable residues R

$$H_2N-\underset{\underset{R}{|}}{\overset{\overset{H}{|}}{C}}-CO-NH-\underset{\underset{R}{|}}{\overset{\overset{H}{|}}{C}}-CO-NH-\underset{\underset{R}{|}}{\overset{\overset{H}{|}}{C}}-COOH$$

3.3.1 Initiation of Translation

To initiate the translation of the mature mRNA molecule into a nascent polypeptide chain, the mRNA (as a substrate of this reaction) has to be recognized first. Detection happens through the poly-A binding proteins at the 3' end of the mRNA. The 5' cap of the mRNA is bound by the eukaryotic initiation factor 4E (eIF-4E), which cooperatively interacts with another one of these factors called eIF4G. Next, a carrier is required to link the mRNA sequence to the encoded amino acids. Every three nucleotides form a triplet, which determines the corresponding amino acid (see Sect. 2.3). Every triplet on the mRNA molecule encoding an amino acid is referred to as a **codon**. This carrier, that transfers the amino acid to the correct codon, is called a transfer RNA (tRNA). The tRNA contains an RNA stretch complementary to the codon, called the **anti-codon**. Depending on the anti-codon, the tRNA is loaded with the amino acid encoded by the corresponding complementary codon (Fig. 3.6). The first codon of an mRNA molecule is called the start codon and always stands for methionine. To initiate translation, a ternary complex (i.e. with three components) needs to form. This complex must contain the eukaryotic initiation factor eIF2, the initiator tRNA and the methionine loaded on that tRNA. The place where the translation is carried out is the ribosome.

Fig. 3.6 The tRNA contains an anti-codon, which is complementary to the codon on the mRNA. Depending on these triplets, the tRNA is loaded with the corresponding amino acid

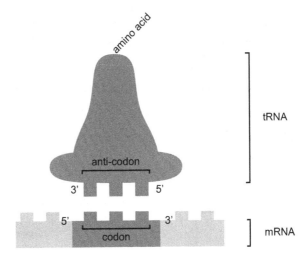

3.3.2   Ribosomal Structure and Function

Ribosomes are modular structures composed of a small and a large subunits, which differ in size between bacteria and eukaryotes. The size of the ribosomes and their subunits was indirectly measured by a centrifugation gradient with the unit S(vedberg) corresponding to the sedimentation of the molecules that does not scale linearly with their size. With respect to the measured sedimentation, ribosomes and their subunits were named accordingly:

Ribosome	Bacteria	Eukaryotes
Small subunit	30S	40S
Large subunit	50S	60S
Total:	70S	80S

In eukaryotes, the ternary complex, harbouring the initiator tRNA loaded with methionine for the start codon at the mRNA bind to the 40S (i.e. small) ribosomal subunit, thereby completing the assembling of the pre-initiation complex. The interaction between eIF4E, which is bound to the 5' cap of the mRNA, and eIF4G, which is bound to the poly-A binding proteins at the 3' end, causes the mRNA to form a circle. The loop formation ensures that the mRNA was correctly processed, otherwise, eIF4E and eIF4G could not interact. In addition, the circular structure facilitates the recruitment of many ribosomes. An ensemble of *many* ribosomes is referred to as *poly*somes.

The small subunits of the assembled ribosomes are already equipped with the initiator tRNA and the methionine. Methionine is always the first amino acid of the translated polypeptide chain, but it will be cleaved by a protease (an enzyme that can degrade proteins). The assembled ribosome moves along the mRNA in their 5' → 3' direction (i.e. from the cap towards the poly-A tail). The ribosome exhibits three distinct sites where codons of the mRNA strand are placed and tRNAs loaded with corresponding amino acids are recruited. These sites are called the A(cceptor), the P(eptide bond formation) and the E(xit) site, respectively (Fig. 3.7).

Fig. 3.7 The assembled ribosome contains three distinct sites where subsequent codons of the mRNA are placed: The A(cceptor), the P(eptide bond formation) and the E(xit) site

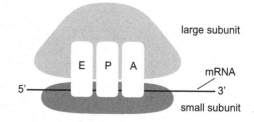

The polypeptide chain is assembled in a four-step process:

1. The specific tRNA with the anti-codon complementary to the codon of the mRNA is recruited to the vacant A site of the ribosome. The tRNA is loaded with the corresponding amino acid.
2. A peptide bond is formed between the amino acid of the tRNA at the A site and the amino acid (chain) of the tRNA at the P site of the ribosome.
3. The ribosome translocates so that the tRNA, which has previously been at the P site, is now at the exit site while the tRNA at the former A site is now at the P site and carries the polypeptide chain.
4. The tRNA at the E site of the ribosome is released and the cycle can continue to assemble the polypeptide chain starting from the N-terminus to the C-terminus.

Similar to the hybridization between DNA and RNA nucleotides during transcription catalysed by the RNA polymerase, also the ribosome exhibits kinetic proof-reading activity. If the tRNA with the wrong anti-codon and amino acid binds to the codon of the mRNA, the ribosome slows down. This deceleration increases the chances that the wrong tRNA falls off the A site and is replaced by the correct tRNA with the anti-codon complementary to the codon, which is energetically favoured. During translation, the mRNA template is already de-capped and de-adenylated, in other words: The 5' cap and the poly-A tail are removed to accelerate the breakdown of the mRNA into individual nucleotides and their recycling to the next round of transcription.

Translation is terminated at the last and so-called stop codon. Once the A site of the ribosome reaches a stop codon, termination factors are recruited, the ribosomal subunits fall apart and the polypeptide chain is released. The region between the start codon and the stop codon is called "open reading frame".

3.3.3 Protein Folding

The synthesized polypeptide chain is at first a one-dimensional sequence of amino acids. To exert its function as a protein, the molecule needs to fold into a 3D structure. The problem is that the angle of the bond between the amino group and the central carbon atom as well as the angle of the bond between the central carbon atom and the carboxyl group can rotate almost freely. As a consequence in theory 10^{30} possibilities exist to fold a polypeptide sequence that consists of 100 amino acids into a 3D structure. Folding took almost infinitely long if all possible conformation would be randomly tested until the correct one was found. In fact, polypeptide chains fold into their 3D structure quite rapidly, once they are released from the ribosome. This discrepancy between the theoretically long folding time and the factual quick folding is known as Levinthal's paradox. The cell contains other proteins, which assist polypeptide folding. These so-called chaperones can also trigger stress responses to clear incorrectly folded proteins thereby preventing their aggregation and further harm (see Sect. 5.3.1).

3.4   Measuring Cellular Content

For the first time, modern biology allows to experimentally assess gene expression on a global scale. This means that all genes can be measured at once, either at the genome level of DNA, or the transcriptome level of mRNA, or the proteome level of proteins and their modifications. Previous studies focused on the investigation of individual genes, their DNA sequence, mRNA abundance or protein level. Now, the whole entity of DNA, RNA or protein can be measured even quantitatively, with absolute numbers of DNA sequence variants, mRNA molecules or proteins per gene. Thus, every instance of gene expression (c.f. Fig. 3.1) can be assessed globally. These measurements of the gen*ome*, the transcript*ome* or the prote*ome* are referred to as the **omics**. Despite the steady development of new methods and improvement of existing technologies, limitations have to be considered and data should be analysed with great care.

3.4.1   DNA Sequence Variants

"Reading" (i.e. measuring) of the DNA sequence is referred to as DNA sequencing. Due to improvements in technology, costs of DNA sequencing dropped significantly in the last years: https://www.genome.gov/sequencingcosts. Since 2008, the reductions in DNA sequencing costs even outpace Moore's Law, a well-established and widely applied model for the development of technology based on the assumption (among others) that computer power doubles every 2 years. Improvements in sensitivity and cost-efficiency of DNA sequencing methods accelerated research findings in the field of genomics. Increasing complexity and diversity of sequenced genomes has been witnessed thanks to the next-generation sequencing technologies [40].

DNA is sequenced to access the genotype of an organism (c.f. Fig. 3.1). The genotype can vary between individuals as they acquire mutations. The word **mutation** is derived from the Latin verb *mutare*, which means "to change". The DNA sequence of a mutant has changed compared to the "wild-type", which serves as a null reference. At a larger scale on the level of chromosomes, changes to the DNA can be either insertions or deletions. Insertions can be subdivided into inversions (i.e. flip of two chromosomal regions), duplications (i.e. copy and paste of a chromosomal region) or additions of genetic material (Fig. 3.8). An example of the latter one is the addition of a DNA region of the human chromosome number 9 to the chromosome number 22. This "translocation" of a chromosomal region creates a fusion gene: BCR-ABL1, a mutation strongly associated with cancer formation [91]. It is apparent that changes in the chromosome level can have drastic consequences for a cell or the entire organism.

At a smaller scale on the level of individual nucleotides, the consequences of mutations are constrained by the fact that the genetic code is degenerated (see Sect. 2.3). It can well be that a change of a single nucleotide does not have any consequences because the mutated triplet also encodes for the same amino acid as the wild type does. If this is the case, the mutation is referred to as "silent". If the mutations of

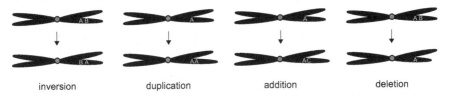

inversion   duplication   addition   deletion

Fig. 3.8 Mutations on the level of chromosomal regions A, B and C are either insertions (inversion, duplication or addition) or deletions

a single nucleotide generate a triplet that encodes for a different amino acid, it is referred to as "missense", because the sense of the sequence might change. The mutant amino acid will have a different spatial arrangement and maybe even other properties (e.g. hydrophilic/-phobic) than the wild-type amino acid, which will influence the 3D structure and thus the function of the polypeptide chain. The most drastic case is the creation of a stop codon through the mutation of a single nucleotide. This mutation will cause the termination of translation at this point and create a fragmented polypeptide chain. The creation of a stop codon is therefore referred to as a "nonsense" mutation. If there is only a single nucleotide (ex)change, the mutation is referred to as "point mutation" (Table 3.1). A change in a single nucleotide can cause many (Greek: *poly*) differently shaped (Greek: *morphos*) organisms. Another name for these mutations is therefore single-nucleotide polymorphism (SNP). It is hard to prove that a mutation in a single nucleotide is solely responsible for changes in the phenotype. That is why genome-wide associations are studied only to find loose connections between SNPs and a disease. The discovered associations need to be validated mechanistically. An example are the mutations of a single nucleotide in the genes MLH1, PMS2 and MSH5. Point mutations in these genes seem to be risk factors for male infertility [50]. In this case, the polymorphism means that the wild type exhibits normal sperm, whereas the mutant exhibits crippled sperm.

Mutations at the nucleotide level can involve insertions or deletions that affect more than a single nucleotide. Collectively, insertions and deletions are abbreviated "indels". If the number of inserted or deleted nucleotides is not a multiple of three, the reading frame will be shifted, because the genetic code is processed in triplets (see Sect. 2.3). A frame shift will alter all encoded amino acids downstream of the

Table 3.1 Point mutations (in grey) of single nucleotides at the DNA level and their consequences on the RNA level of the codon and the amino acid level of the protein. Missense mutation to lysine is very similar to wild-type arginine but different from threonine

	Wild-type	Mutation			
		Silent	Missense		Nonsense
DNA	TCT	TCC	TTT	TGT	ACT
RNA	AGA	AGG	AAA	ACA	UGA
Protein	Arginine	Arginine	Lysine	Threonine	Stop codon

mutation. In the following example, the deletion of a single nucleotide altered already the start codon, inserted a stop codon and changed all other encoded amino acids:

wild-type codons	AUG	CAU	UUG	AGA	AGG	ACA	UGA
wild-type amino acids	M	H	L	R	R	T	Z
mutant codons	AGC	AUU	UGA	GAA	GGA	CAU	GA
mutant amino acids	S	I	Z	E	G	H	

All such variants can be measured by DNA sequencing. If the sequencing is not "deep" enough, some sequence stretches will not be detected. Count matrices are obtained (Table 3.2), which can be analysed for statistical significance of the presence or absence of certain mutations under different conditions (e.g. healthy versus disease condition).

3.4.2 Omics Data

Independent of the platform and its specifications during the experiment and the measurement, processed genome, transcriptome and proteome data have the same format. Per genetic region and condition, obtained levels of sequence variants (i.e. DNA), transcripts (i.e. mRNA) or protein are given in absolute or relative units (Table 3.2).

Recent advances in transcriptomics allow the rapid and standardized acquisition of mRNA numbers per gene (i.e. protein-coding genetic regions), with every condition being a single cell. Why is it important to measure the mRNA globally at the resolution of single cells? Because the cellular composition of a tissue can be a heterogeneous mix of different cell (sub)populations, which cannot be discriminated when being averaged out in a bulk measurement. Single-cell RNA sequencing can discover rare cell types, which have to be characterized based on their transcriptome. The identification of a unique set of marker genes that distinguish one subset of cells from another one is a non-trivial task. Recent technological advances allow to resolve heterogeneity in a spatial context including physical interactions for the improved assessment of cellular identity and biological function [38]. Computational pipelines help to understand that the acquired data represents snapshots of dynamic temporal processes [2], e.g. the progression of a disease such as liver fibrosis.

Table 3.2 For genomics, transcriptomics and proteomics processed output is displayed as a counts table for genetic regions under different experimental conditions

	Condition 1	Condition 2	...	Condition n
Genetic region 1	32	NA	...	1
Genetic region 2	NA	15	...	NA
...	...	...	...	...
Genetic region n	6	NA	...	7

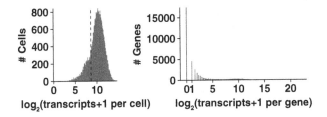

Fig. 3.9 Distributions of the numbers of measured transcripts per cell (left) and per gene (right). Dashed vertical line indicates cut-off (i.e. 400 transcripts) for quality control. Data from [49]

The obtained data matrices (c.f. Table 3.2) are sparse with missing entries for many gene(tic region)s/conditions. Therefore these omics data require proper processing. Raw data has to be inspected carefully to learn something about their distribution (Fig. 3.9) for quality control. A normalization procedure with proper controls needs to be established to compare datasets from different experiments. Because both, the transcriptome and the proteome, are intricately regulated at several layers, global trends in gene expression are non-intuitive. Ideally, published findings are freely available as raw data as well as processed data together with curated open-source code to reproduce results sustainably.

An example is the data in [49], which can be inspected and (re)analysed with all the information provided on a repository under: http://bit.ly/2JvmlD7.

Summary

- For transcription of DNA into mRNA, conserved sequences have to be bound by transcription factor molecules.
- Transcription is terminated and transcripts are processed by splicing: Introns are cut out, while exons are being pasted together.
- mRNA is translated by ribosomes into a polypeptide chain, which needs to be folded and shuttled to the correct location.
- Total DNA, mRNA or protein content of cells can be measured with current-gen/transcript/prote"-omics" technologies.

Regulation of Gene Expression

4

Contents

Introduction

The mechanisms of gene expression (i.e. the decoding of genetic information as a blueprint to synthesize proteins with a certain structure and function) are conserved between bacteria and eukaryotes. With additional complexity scales the possibility for regulatory layers to fine-tune processes during gene expression. In biology, structure and function are tightly coupled. The composition of DNA, RNA and protein allow interactions among themselves and each other. Interactions lead to the creation of regulatory networks rather than a linear chain from DNA to mRNA to protein (Fig. 4.1). These networks evolved with time and allowed adaptation to changing conditions. If more structural proteins are needed to allow the rapid development of more complex cellular structures and multi-cellular tissues of an organism, all steps from DNA transcription to RNA translation and protein folding need to be balanced to prevent bottlenecks in the production. An increased pool of mRNA might facilitate rapid translation once the respective protein is needed. Therefore, micro-RNA (miRNA) exist, which can for instance block translation without necessarily affecting the production of mRNA. On the other side, miRNA can lead to the degradation of mRNA, which needs to be precisely balanced.

Dynamic synthesis and degradation, as well as the possibility for acceleration or deceleration at every step of gene expression, can save cellular resources. Because a protein does not have to be produced from an incorrectly processed mRNA. Instead,

© Springer-Verlag GmbH Germany, part of Springer Nature 2022
L. Adlung, *Cell and Molecular Biology for Non-Biologists*,
https://doi.org/10.1007/978-3-662-65357-9_4

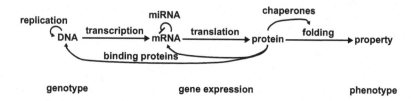

Fig. 4.1 The concept of gene expression does not follow a linear process but contains several feedback and feedforward mechanisms resulting in a complex regulatory network

the process can be abrogated as soon as an error occurs. Maintenance of such regulatory mechanisms also requires energy and metabolites but allows the cell more valuable flexibility when facing a change in conditions.

4.1   Transcriptional Regulation

While many paradigms were disproved in the light of modern technologies, the key concepts for regulation of gene expression dating back to the work of Jacob and Monod in the 1960s do still hold [48]. From their pioneering work in bacteria, they proposed the existence of proteins, which they called *trans*-factors, that are capable of binding DNA at specific sites, which they called *cis*-elements. When *trans*-factors bind to *cis*-elements, they recruit the molecular machinery to initiate transcription. The *trans*-factors are meanwhile called **transcription factors**. In eukaryotic cells, these transcriptional control mechanisms were extensively studied, too. Today's common notion is that transcription factors bind to DNA, recruit cofactors and the transcription apparatus as well as chromatin remodelling enzymes to regulate gene expression on the transcriptional level.

In eukaryotes, transcription factors cannot only bind to the promoter region of a gene but also to the distal (i.e. far away) sites upstream of the transcription start site. The distal DNA regions, bound by transcription factors, are called **enhancers**, because they enhance the recruitment of RNA polymerase II. The presence of the general transcription factors TFIIB, TFIID, TFIIE and TFIIH in addition to RNA polymerase II is sufficient to initiate transcription (see Sect. 3.1.1). To bring distal regulatory sites and their transcription factors in close proximity to the transcription start site, the so-called "Mediator" complex is required, too. Binding of the mediator to transcription factors and RNA polymerase II leads to a bending of the DNA and helps to efficiently launch transcription (Fig. 4.2). As the organization of the eukaryotic genome is complex, a chromatin-remodelling complex is also part of the transcription machinery to ensure open chromatin and accessible DNA. Remodelling of chromatin is achieved by histone-modifying enzymes (see Sect. 4.4).

There are also regulatory mechanisms that inhibit the initiation of transcription. GC(-rich) boxes at the DNA sequence tend to be chemically modified by the addition

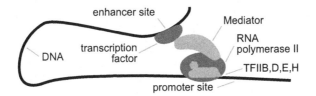

Fig. 4.2 The mediator complex bridges the RNA polymerase II and the transcription initiation factors TFIIB, D, E and TFIIH at the promoter site with a transcription factor at the enhancer site of the DNA

of a methyl group. These methylated nucleotides impede the binding of transcription factors and, instead, recruit histone-modifying enzymes that create heterochromatin, which cannot be transcribed and is thus considered "transcriptionally silenced".

4.2   Regulatory RNAs

As transcription is a very essential cellular process, its misregulation is closely linked to diseases ranging from cancer (see Chap. 9) to autoimmune disease, diabetes and developmental disorders [68]. Mutations that cause such deleterious consequences can either affect the genes encoding for a transcription factor or their DNA-binding sites. The latter ones are the *cis*-elements, which can be either located at the core promoter or distal sites.

The complexity of an organism does neither correlate with DNA Content nor with the number of protein-coding Genes, which is known as the C-value paradox and the G-value paradox, respectively. For instance, the genome size of common wheat *Triticum aestivum* exceeds with 16×10^9 base pairs the size of the human genome with 3×10^9 base pairs by far. Despite its reduced complexity compared to humans, the roundworm *Caenorhabditis elegans* contains with $\approx 20,000$ the same number of protein-coding genes as *Homo sapiens*. These discrepancies can be resolved when looking at the number of regulatory elements in the genome, because the number of non-protein-coding sequences per genome scales with the organismal complexity [100].

Among the RNA molecules transcribed from DNA, which do not encode for proteins, the two most prominent are:

- long non-coding RNA and
- micro-RNA (miRNA).

Both types of these non-coding RNA are synthesized mainly by RNA polymerase II. While the function of the majority of long non-coding RNAs is not known, the role of **miRNAs** as an important regulatory player in gene expression is well understood. After being transcribed, miRNAs exist as a pri-miRNA usually in the shape of a hairpin (c.f. Fig. 3.3). This pri-miRNA is then shortened by an enzyme called Drosha, which releases a pre-miRNA. The pre-miRNA is subsequently exported from the nucleus to the cytoplasm. The cytoplasmic pre-miRNA is cleaved by another enzyme called DICER. The resulting double-stranded miRNA (called "miRNA duplex") is separated into single strands of mature miRNA, each ≈ 21 nucleotides long, by a helicase. The mature miRNA is loaded onto a protein called Ago to form the RNA-induced silencing complex (RISC). Silencing (i.e. shutting down gene expression) can happen in two ways:

1. Perfect match of the miRNA on the RISC with a recruited mRNA causing the degradation of the latter.
2. Imperfect binding of the miRNA on the RISC with an mRNA causes a block of ribosome binding and thus inhibits translation initiation.

The first scenario almost entirely *abrogates* the expression of the gene transcribed into mRNA that matches the miRNA perfectly: $\Rightarrow$ 0% protein.

The second scenario slows down translation but keeps the mRNA intact and thus only *reduces* the expression of the respective gene: $\Rightarrow$ [20%; 80%] protein.

Either way, regulatory RNAs interfere with the initiation of translation.

4.3   Protein Control

Once translation is successfully initiated, protein synthesis (i.e. the translation of mRNA into polypeptide chains) is mainly controlled through the translation rate, in other words, the speed of the ribosome sliding along the mRNA. Mostly, the ribosome is physically blocked and therefore halts in its progression. An example is the control of iron metabolism. If a cell is being starved from iron (i.e. $Fe^{2+/3+}$ ions), it needs less ferritin, a molecule that can bind free iron, but more of the transferrin receptor, which facilitates the import of iron into the cell. In case of low cellular iron levels, aconitase is bound to an mRNA hairpin structure in front of the ferritin mRNA and behind the transferrin receptor mRNA. Thus, only the transferrin receptor mRNA is translated, and more transferrin receptor is produced, which in turn accelerates the iron uptake into the cell. In case of iron access, the $Fe^{2+/3+}$ ions bind to aconitase and cause a conformational change in the protein and its release from the mRNA. As a consequence, the ferritin mRNA can be translated to produce more ferritin and bind free iron, while the mRNA of the transferrin receptor is degraded (Table 4.1). This fine balance is kept through selective blockade of the ribosome.

4.3.1   Post-Translational Modifications

Once the polypeptide chain has been synthesized and the mature protein folded correctly, its activity can be controlled by chemically modifying it. Since these modifications of the protein occur after the translation of the mRNA, they are referred to as "post-translational modifications". Modifications are **enzymatic reactions**, which facilitate the addition or the subtraction of groups of molecules.

Table 4.1 Cellular iron metabolism is regulated by the control of translation. The more protein of the transferrin receptor (CD71) is synthesized, the more of the missing $Fe^{2+/3+}$ ions can be imported into the cell. The more of the ferritin protein is present, the more of the excessive iron can be bound

$Fe^{2+/3+}$	Aconitase	Ferritin mRNA	CD71 mRNA	Ferritin protein	CD71 protein
Starvation	Bound to hairpin	Blocked	Protected	⇓	⇑
Excess	Bound to iron	Translated	Degraded	⇑	⇓

An enzyme (E) is a protein converting a substrate molecule (S) thereby creating reaction product (P), with an intermediate state of an enzyme-substrate complex (ES). The enzyme as a molecular machine is not needed for the reaction itself ($\rightarrow$) to occur, but it speeds up ("catalyses") the reaction:

$$E + S \rightleftharpoons ES \rightarrow E + P.$$

The velocity v, which represents the reaction rate, can be expressed by the following ordinary differential equation:

$$v = \frac{d\,[P]}{dt} \Rightarrow \frac{V_{max} \cdot [S]}{K_M + [S]}, \tag{4.1}$$

with $d\,[P]/dt$ describing the change d of the concentration of the product [P] with the change d of time t. [S] denotes the concentration of the substrate S. V_{max} denotes the maximum velocity of a reaction assuming saturation with substrate. The parameter K_M is known as the Michaelis-Menten constant, which specifies the substrate concentration at which the reaction rate is half-maximal. This becomes evident if K_M in Eq. 4.1 is substituted with [S]:

$$v|_{K_M=[S]} = \frac{V_{max} \cdot [S]}{[S] + [S]} \Rightarrow \frac{V_{max} \cdot 1}{1 + 1} \Rightarrow \frac{1}{2} \cdot V_{max}. \tag{4.2}$$

The concentration of the enzyme [E] is not even part of the equation because the substrate concentration [S] is the rate-limiting step. The molecular machinery of the enzyme E works so efficient that very little amounts are saturating. A dose-response curve of an enzymatic reaction showing the velocity of the reaction v depending on the concentration of the substrate [S] resembles a hyperbolic function (Fig. 4.3).

An example of an enzymatic reaction is the hydrolysis of guanosine triphosphate (GTP), which is the substrate S of the enzyme, to guanosine diphosphate (GDP) and phosphate, which are the products P of this reaction (Fig. 4.4). The enzyme catalysing the *hydrolysis* of GTP is a *hydrolase*; a *GTPase* to be precise. GTPases are fundamental enzymes for the regulation of cellular processes as it will be described in Sect. 5.2.1.

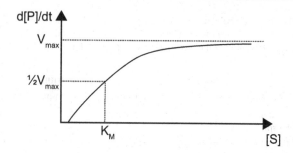

Fig. 4.3 Dose-response curve showing velocity of the reaction $v = d\,[P]\,/dt$ depending on the concentration of the substrate [S], with K_M being the substrate concentration of the half-maximal velocity of the reaction

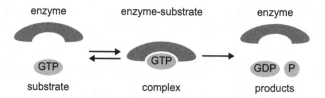

Fig. 4.4 Enzymatic reaction of a GTPase enzyme, forming a complex with its substrate GTP to hydrolyse it and release GDP and a phosphate group (P) as a product

Modifying residues of proteins through enzymatic reaction is a way to control their function without changing their absolute amounts, even though post-translational modifications can lead to a direct change in absolute amounts as is the case for targeted protein decay.

4.3.2   Protein Decay

Protein levels cannot only be controlled by the production rate through the regulation of transcription and translation, but absolute protein abundance can also be controlled through targeted degradation. While RNA molecules are susceptible to hydrolysis and thus decay by themselves with a certain half-life, proteins need to be broken down actively. Proteins are labelled with small molecules, which flag them for degradation. The most prominent mark for protein degradation is the small molecule **ubiquitin**. Its name is derived from the fact, that the molecule is "*ubiquitously*" found in almost every cell and every tissue. Protein degradation through ubiquitination is tightly regulated. First, proteins subjected to degradation are labelled with a linear chain of ≥ 4 ubiquitin molecules. This degradation flag is recognized by a cellular machinery called the 26S proteasome - the cellular protein dumpster. Similar to the ribosomes (see Sect. 3.3.2), the size of the proteasome is indicated with the unit S(vedberg). The 26S proteasome consists of a central core unit and two regulatory caps at the top and the bottom of the core (Fig. 4.5). The ubiquitin-tagged proteins are unfolded,

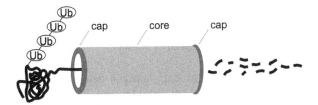

Fig. 4.5 Proteins labelled with a linear chain of at least four ubiquitin (Ub) molecules are being degraded into small fragments inside the 26S proteasome

which requires energy from ATP hydrolysis. The polypeptide chain is subsequently pulled into the core of the 26S proteasome and chopped into small fragments of seven to nine amino acids in length. These fragments and the ubiquitin molecules are recycled.

Degrading proteins prevents their accumulation and helps to keep a protein balance (i.e. "protein homoeostasis" or "proteostasis") in the cell. However, the logic behind this regulation is more complex, as there are proteins, which are labelled with ubiquitin, but they will *not* be degraded, and there are proteins, which are *not* labelled with ubiquitin, but they will be degraded [26]. The dynamic production and degradation of proteins leads to the fact that protein content is not inherited from one cell to another throughout generations. If a cell divides to form two daughter cells, proteins of the mother cell are distributed among the daughter cells (see Sect. 7.3.2), but the half-life of a protein is usually much shorter than the generation time (i.e. the duration between two cell divisions).

4.4 Epigenetics

On top of genetic information encoded in DNA that will be maintained and inherited from one generation to another, there are chromatin states that determine whether genes are accessible and can be expressed, or not. This additional layer above the genetic code is referred to as epigenetics (from Greek *epi* as prefix meaning "over" or "on top of"). This means the epigenetic information is an additional layer on top of the inheritable DNA sequence, which describe the chromatin state within a cell. Despite adding a regulatory layer on top of the DNA, the chromatin state is not independent of the DNA sequence.

4.4.1 Mechanisms of Epigenetic Regulation

Different factors are involved in the regulation of the chromatin state. The molecular complexes introducing modifications can be classified as:

- writers—creating modification marks,
- readers—interpreting written modification marks or
- erasers—removing written modification marks.

The local change of the chromatin state between hetero- and euchromatin (see Sect. 2.2.1) is a process called remodelling and means the structural rearrangement between closed and open chromatin.

Whether chromatin is tightly or loosely packed depends mainly on the chemical modifications of the DNA itself, and of histone proteins, around which the DNA is wrapped. Common modifications at the DNA level include the addition of methyl groups in particular to cytosine bases. Modifications of histone proteins mainly involve the addition of methyl or acetyl groups or a combination of those.

At the DNA level, methyl groups are transferred to cytosine bases by DNA methyltransferase (DNMT) enzymes. DNMT enzymes are thus the **writers** of DNA methylation. The cytosine base of a $\frac{C}{G}$ nucleotide pair is often referred to as "CpG". Methylated cytosine bases can be bound by methyl-CpG-binding domain (MBD) proteins. Those MBD proteins are **readers** of DNA methylation. Once bound to methylated DNA, MBD proteins recruit other proteins that can modify the chromatin structure. These "chromatin remodelling factors" induce locally the formation of heterochromatin, which is closed and does not allow for gene expression. Thus, methylated cytosine bases, particularly in the proximity of promoter sequences, commonly lead to the silencing of gene expression.

Removal of methyl groups from cytosine bases mostly occurs passively, when DNMT enzymes are not active. Methylated cytosine bases can be converted to thymine bases by spontaneous deamination (i.e. the removal of an amino group: $-NH_2$). This conversion creates a $\frac{T}{G}$ nucleotide pair from the original, correct $\frac{C}{G}$ nucleotide pair. This erroneous conversion of methylated cytosine to thymine in CpG nucleotide pairs can be corrected by "base excision-repair" (see Sect. 2.4.2). Thus, there are no directly active **erasers** of DNA methylation (i.e. DNA demethylases). Among the chromatin remodelling factors that are recruited to the DNA in the presence of DNA methylation and MBD proteins are histone-modifying enzymes (Fig. 4.6).

A prime example of how histone modifications regulate gene expression is the silencing of the inactive X chromosome [45], through the addition of three methyl groups at the Lysine residue number 27 at the histone 3 ("H3K27me3"). Histone-

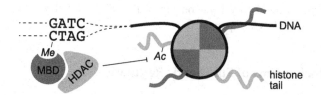

Fig. 4.6 Methylated (Me) cytosine residues in the DNA are recognized by methyl-binding domain (MBD) proteins, which in turn recruit histone deacetylase (HDAC) enzymes that remove acetyl (Ac) groups from histone tails and repress gene expression

Table 4.2 Histone modifications by respective enzymes

Histone modification	Enzyme	Class
Acetylation	Histone acetyltransferase	Writer
Deacetylation	Histone deacetylase	Eraser
Methylation	Histone methyltransferase	Writer
Demethylation	Histone demethylase	Eraser

modifying enzymes can be recruited upon the change of the methylation pattern of DNA nucleotides. Addition of methyl groups directly to nucleotides in the DNA sequence thus represents another epigenetic layer highlighting the complexity of regulatory mechanisms of gene expression.

Modifications of histone proteins are introduced by **writers** of the different chemical residues, which are added to the histone tails, and removed in the reverse reaction by **erasers** of histone modifications (Table 4.2).

Depending on the chemical properties of the added modifications (e.g. electrical charge), the histone-DNA assembly can either be disrupted (by repulsive forces) or tightened (by attractive forces). For example, the addition of a negatively charged acetyl group to a histone tail through a histone acetyltransferase (HAT) enzyme leads to the repulsion of the negatively charged backbone of the DNA (see Sect. 2.1), because like charges repel each other. Consequently, histone acetylation induces unwrapping of the DNA from the acetylated histone protein, which in turn allows for gene expression from the locally formed euchromatin. The reverse reaction is catalysed by HDAC enzymes removing acetyl groups from histone tails. The remaining unmodified lysine residues of the histone tail are positively charged and thus attract the negatively charged backbone of the DNA. The tight wrapping of histone and DNA in this locally induced heterochromatin conformation does not allow for gene expression.

Due to the combinatorial complexity of modified histone residues, it is hard to predict the effect of a single histone modification on the recruitment of chromatin remodelling factors and gene expression. The functional consequences of the modification of a single histone residue depend on the presence or absence of other histone modifications. The overall pattern of all present histone modifications is referred to as the **histone code**, which influences chromatin structure and function by creating or removing binding sites for other chromatin remodelling factors.

These chromatin remodelling factors are the **readers** of histone modifications. E.g. acetylated Lysine residues of histone tails are recognized by the bromodomain of transcription factors (see Sect. 4.1), which activate gene expression. This example shows that the effect of the chromatin remodelling factors (i.e. recruitment of transcription factors → activation of gene expression) can oppose the physicochemical properties of the histone modification itself (i.e. negative charge of acetylation → deactivation of gene expression), which introduces an additional layer of epigenetic regulation.

4.4.2 Causes and Consequences of Epigenetic Regulation

The link from the DNA sequence to the property of an encoded protein (see Fig. 4.1) can be modulated through mechanisms of epigenetic regulation. Therefore, the connection between a genotype and a phenotype (see Sect. 2.2.1) is ambiguous.

Organisms with the same genetic information (i.e. the same DNA sequence) can respond differently, e.g. when being exposed to different environmental conditions. In some turtles, the sex of the progeny is determined by the temperature surrounding the egg. In this case, an environmental factor (i.e. temperature) influences the demethylation of, again, the three methyl groups at Lysine residue number 27 on histone 3 ("H3K27me3"), which in turn regulates the expression of the sex-determining gene *Dmrt1*. Two genetically identical eggs can give rise to male progeny at 26 °C or female progeny at 32 °C, due to epigenetic regulation induced by the environment [37].

This example shows that epigenetic regulations play a prominent role in the development of higher organisms. Unless mutations are introduced, all our body cells contain the same DNA sequence. The differences in cellular identity originate from the fact which genetic information is expressed, and which genetic information is repressed—either through chromatin states or regulatory RNAs [31].

Evidently, the regulation of gene expression at different levels adds plasticity to the possible programs exerted by the genetic code, which gave rise to complexity and diversity during both, development and evolution.

Summary

- Regulatory RNA molecules can directly interfere with gene expression, e.g. by binding of miRNA to mRNA, inhibiting their translation or inducing their degradation.
- Proteins are extensively modified upon translation, which influences their activity and stability.
- Epigenetic modifications occur on the level of DNA (e.g. methylation) and histones (e.g. acetylation).

Membranes and Intracellular Transport

5

Contents

Introduction

All reactions that occur within a cell can proceed more efficiently if they are spatially separated from each other. Emergence of life from a self-replicating compartment required, besides the hereditary material, a reaction chamber. The walls of this reaction chamber are membranes made out of lipids. But these membranes do not only separate the intracellular reactions from the exterior but also wrap cellular cargo in vesicles for shipping within the cell or secretion to the outside (Fig. 5.1). As both, plasma membranes and cellular vesicles, are composed of lipids, they can be regulated together. To secrete the material they carry to the extracellular space, vesicles can be fused with the plasma membrane. The reverse reaction takes place, too, if vesicles are budding from cellular membranes to take up cargo for transport and processing.

There is a dynamic exchange between the outside and the inside of a cell in response to the environment. Intracellular transport demands frequent re-organization of membranes. The adaptation of cellular compartment structures to fuel reactions and production chains is essential for the cell to survive. The creation, merging and budding of membranes are made possible by their biochemical composition. Lipids are amphiphilic molecules, which means that, in part, they prefer both, aqueous and non-aqueous solutions. Therefore, they assemble into bilayers (Fig. 5.1) to protect their hydrophobic part from water and expose the hydrophilic part to the surrounding aqueous solution. Regulatory components are additionally required to orchestrate membrane exchange and intracellular trafficking. These processes are an intriguing subject because they highlight once more that complex biological systems

© Springer-Verlag GmbH Germany, part of Springer Nature 2022   49
L. Adlung, *Cell and Molecular Biology for Non-Biologists*,
https://doi.org/10.1007/978-3-662-65357-9_5

Fig. 5.1 Lipid bilayers form
and separate cellular exterior
from the interior thereby
creating reaction
compartments

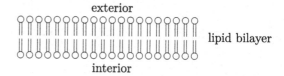

exterior

lipid bilayer

interior

can be assembled from rather simple building blocks yielding sophisticated molecular apparatuses such as the endoplasmatic reticulum (ER) made out of lipids for specific modifications of proteins.

5.1   Composition and Assembly of Cellular Membranes

The cell is a biochemical container full of specific molecules reacting together. There are separating borders to prevent reactions of these components with molecules of separate compartments or the outside medium. These reaction barriers are composed of lipids in a so-called bilayer. Two ply sheets of lipids are penetrated by proteins to allow the targeted import and export of molecules for information exchange. The composition and dynamic assembly of cellular membranes is a vital task that is carried out by all living cells. The plasma membrane needs to be continuous and elastic to enable integrity despite deformations and growth, which is achieved by the biochemical properties of the membrane components.

Lipids consist of a polar head, which is therefore soluble in water. Phospholipids such as phosphatidylcholine are a common membrane component. The phosphate group at the head of phospholipid is negatively charged and forms electrostatic attractions with water molecules and is thus considered "hydrophilic". This water-loving side of the lipids is opposed by a pair of long water-fearing tails. These carbon chains are neither charged nor polar and therefore not loving water but "hydrophobic". If lipids are put into an aqueous solution they will arrange in the energetically most favourable way. This means the polar head will be exposed to water, whereas the uncharged tail will shun water. The tails will form an inner core of two opposed sheets of lipids with an outer shell of polar heads. To prevent exposure of water to the tails at the edge of this lipid bilayer, the sheets have to bend and enclose in a three-dimensional (3D) vesicle (Fig. 5.2).

The bended membrane has to seal its interior. The lipid bilayer has to be flexible without rupturing. Whether a membrane is rather rigid or rather fluid results from the tails of the individual lipids. These chains contain between 14 and 24 carbon atoms. The tails of the most common lipids contain 18–20 lipids. These dimensions build up membranes with an average thickness of 5 nm. If the tails of lipids are shorter, interactions with one another are less likely, which makes the membrane-less rigid. The fluidity is also influenced by the saturation of the carbon atoms in the tails with hydrogen. If all carbon atoms are **saturated**, the chain is rather linear and straight. If two sequential carbon atoms within the tail are **unsaturated**, double bonds are formed between them, which introduces a kink in the chain. Phospholipids

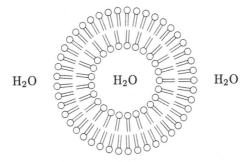

$$H_2O \qquad H_2O \qquad H_2O$$

Fig. 5.2 A droplet-like vesicle with a bilayer shell is the energetically ideal 3D arrangement of lipids in aqueous solution. Polar heads of lipids are exposed to H_2O, while the uncharged tails are separated from water within the bilayer

Table 5.1 Summary of membrane properties. A densely packed membrane is rather rigid, which could be due to long or saturated lipid tails or insertion of many cholesterol molecules. Loosely packed, fluid membranes can occur because of many short or unsaturated lipid tails or only few cholesterol molecules within the bilayer

Packaging	Viscosity	Lipid tails	Bonds	Cholesterol insertions
Densely	Rigid	Long	Saturated	Many
Loosely	Fluid	Short	Unsaturated	Few

with unsaturated and therefore kinky chains cannot be packed as densely as lipids with saturated and thus straight chains. Loosely packed lipids make a membrane more fluid. The same holds true for cholesterol, one of the most common membrane proteins. If **cholesterol** is inserted into the bilayer, it fills lateral gaps between the phospholipids, which makes the membrane stiffer and less flexible. These properties are summarized in Table 5.1.

5.2 Targeted Protein Transport

Lipid membranes form borders to separate compartments for specific reactions within a cell. The nucleus is thought to be evolved from the plasma membrane that turned inside into the cell to surround and separate the DNA. The process of a membrane folding inwards is called invagination. This hypothesis for the creation of cellular organelles is supported by the fact that the nuclear envelope and the ER are composed of an outer and an inner membranes, which assemble through the invagination of the plasma membrane. The space between the two membranes is called the lumen and forms a continuum between nucleus and ER. Since mitochondria and chloroplasts are derived from procaryotes (see Sect. 1.3), parts of their protein synthesis are independent. But in general, most of the proteins are synthesized at ribosomes in the cytoplasm of the cell and need to be dispatched to their correct destination for the maintenance of cellular functions and growth.

The address label is contained in the amino acid sequence of the proteins. It is between 15 and 60 amino acids long and often but not always cleaved upon delivery. The sequence, which directs the protein to the correct compartment, is called the **sorting signal**. The physical properties such as charge and hydrophobicity are more important than the exact sequence of the sorting signal. Sites of protein delivery can be nucleus, mitochondria, chloroplasts, peroxisomes or ER. The protein remains in the cytoplasm in absence of a sorting signal. Proteins destined for the nucleus are delivered through nuclear pores, which are holes in the nuclear envelope. Proteins that have to be transported across membranes to mitochondria, chloroplasts or the ER need to unfold. Specific protein **translocators** facilitate this process. Proteins shipped from the inside of the ER to the Golgi apparatus, lysosomes or endosomes are ferried within **vesicles**.

5.2.1   Protein Transport through Nuclear Pores

The nucleus contains the DNA. For gene expression and its regulation (see Chap. 4), molecules are shuttled in and out of the compartment with high frequency. The nucleus is perforated with **nuclear pore complexes** that span the inner and the outer membranes of the nucleus. More than 1000 translocations occur every second at every single pore. Despite the fast rates, trafficking between cytoplasm and nucleus is specifically regulated. The inner diameter of the pore is $\approx 40\,\text{nm}$, which allows small, water-soluble molecules to pass freely. But larger proteins cannot simply diffuse through the nuclear pore complex [53], which is composed of up to 1000 building blocks, so-called nucleoporins, from ≈ 30 different protein families [47]. Some polypeptide chains of the nucleoporins hang loosely into the inner space of the pore thereby creating a diffusion barrier. Active transport through this meshwork requires energy that is provided by the hydrolysis of GTP. This reaction is catalysed by a small GTPase (see Sect. 4.3.1), named **Ran**. There is a concentration gradient between Ran bound to GTP in the nucleus and Ran bound to guanosine diphosphate (GDP) in the cytoplasm [79].

For a protein to be imported into the nucleus, the required sequence is the **nuclear localization signal**. It is recognized by nuclear import receptors that mediate the way through the fibrils of the nuclear pore complex. The nuclear import receptors bind the protein with the nuclear localization signal and direct it through the pore to the nucleus. Upon delivery of the cargo to the nucleus, the nuclear import receptor binds to a Ran-GTP complex and is thereby shuttled back to the cytoplasm where the GTP bound to Ran is hydrolysed to GDP and phosphate in presence of a GTPase-activating protein (GAP). The resulting high concentration of Ran-GDP in the cytoplasm leads to rapid recycling of the complex to the nucleus, where the GDP bound to Ran is exchanged by another GTP. This reaction is catalysed by a guanine nucleotide exchange factor (GEF) and maintains high amounts of Ran-GTP in the nucleus thereby driving rapid export along the concentration gradient. Ran-GTP is shuttled back from the nucleus into the cytoplasm (Fig. 5.3). Ran-GTP directed from the nucleus to the cytoplasm can also carry proteins along that contain a sequence called

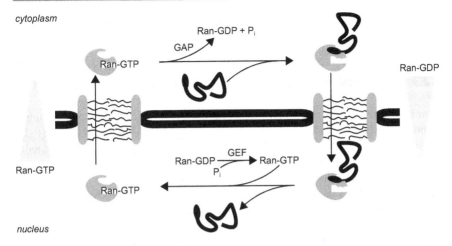

Fig. 5.3 Nuclear import from the cytoplasm. Protein with nuclear localization signal is bound by nuclear import receptor in the cytoplasm and dragged through a nuclear pore complex into the nucleus. After the release of the cargo, the nuclear import receptor binds Ran-GTP. Export of this complex is facilitated by the Ran-GTP concentration gradient

the **nuclear export signal**. Both targeted transport processes, the shuttling from the cytoplasm to the nucleus and the export from the nucleus to the cytoplasm, were recently shown to be light-inducible by biotechnological engineering [81,82].

5.2.2 Unfolding of Proteins for Mitochondria and Chloroplasts

Mitochondria and chloroplasts (see Sect. 1.3) contain their own DNA and thus produce some of their proteins by themselves. However, the majority of mitochondrial and chloroplast genes are transcribed in the nucleus and translated in the cytoplasm of the eukaryotic cells. These proteins need to be imported into the mitochondria or chloroplasts for exertion of their specific tasks and maintenance of the organelle. Unlike the folded proteins entering the nucleus through pores, transport through the outer and inner membrane of mitochondria or chloroplasts requires proteins to be unfolded. The polypeptide chain is directed through **translocator proteins** in the outer and inner membrane like the thread through the eye of a needle. Chaperones pull the imported proteins and help them to fold correctly once they reached the inside.

For proteins to be translocated into mitochondria or chloroplasts, they need a signal sequence, which is recognized by an import receptor. This sequence is cleaved as the destination is reached and the mature protein is correctly folded. Transport to a specific site within the organelle requires an additional sorting signal, which is exposed upon the removal of the first sorting signal. Thereby a polypeptide chain can e.g. be inserted into the inner membrane of mitochondria. Membranes are of vital importance for the functionality of mitochondria and chloroplast. Phospholipids are transported upon synthesis in the ER to proximal membranes of mitochondria and chloroplasts by lipid-carrying proteins that enable distinctive compositions of these membranes.

5.2.3   Co-Translational Import into the Endoplasmatic Reticulum

Proteins destined for the Golgi apparatus, endosomes, lysosomes, the plasma membrane or the cell's exterior are all shuttled through the ER. All proteins that are to be transported there will enter the ER lumen, which is the interior space of the organelle. Transmembrane proteins will be incorporated into the ER membrane. Respective proteins are imported into the ER from the cytoplasm and will never be returning to the cytoplasm along their journey to their target compartment. To be transferred to the ER, proteins need a sorting signal sequence at their N-terminus. The mRNA is translated in 5'-3' direction by ribosomes into nascent polypeptide chains from the N-terminus to the C-terminus so that the sorting signal is synthesized at the beginning. Once exposed, the sorting signal is recognized by a **signal recognition particle (SRP)**. Binding of the SRP to the nascent polypeptide chain slows down translation and drags the complex to the ER where the SRP engages with an **SRP receptor**. Upon this interaction, the polypeptide chain is threaded across the ER membrane through a channel of a translocator protein in the vicinity of the SRP receptor. The translocation occurs while the protein synthesis is resumed by the ribosome at the ER membrane. Co-translational import of nascent polypeptide chains from ribosomes into the ER led to the observation of the **rough ER**, which is membranes covered by ribosomes that produce proteins destined for intracellular trafficking or secretion.

Transmembrane proteins are not fully pulled through the translocator protein in the ER membrane. Instead, they contain a **stop-transfer sequence**, which is hydrophobic and therefore stalls the translocation process. The stop-transfer sequence becomes the membrane-spanning domain that is released from the translocator protein channel laterally into the lipid bilayer of the ER membrane. The N-terminus of this protein is located in the ER lumen, while the C-terminal part remains in the cytoplasm. If a transmembrane protein contains multiple membrane-spanning domains, hydrophobic pairs of ER sorting signal and stop-transfer sequence can be found within the polypeptide chain. Post-translational incorporation of these proteins into the ER membrane occurs similarly to the mechanism described above but without cleavage of the sequences. In a stitching-like mechanism, the hydrophobic regions are dragged into the translocator protein channel within the ER membrane and remain there while the rest of the polypeptide chain is pulled through.

Both, water-soluble molecules ending up in the ER lumen and transmembrane proteins embedded in the lipid bilayer of the ER, can be further transported. As the ER, the Golgi apparatus, lysosomes, peroxisomes and endosomes are forming an interconnected network of membrane-enclosed organelles, they are summarized as the **endomembrane system**. No matter if proteins are secreted from the cell targeted to the lumen of another organelle of the endomembrane system or the plasma membrane, onward-trafficking from the ER involves cellular vesicles.

5.3   Vesicular Transport

Vesicles constitute the intracellular shipping and delivery system. Senders and recipients are always different membranes of the endomembrane system. Cargo is protein within the lumen of the vesicles. Transportation has to be specific in three ways:

1. Only proteins that need to be delivered should be wrapped by a lipid bilayer inside a vesicle.
2. The composition of vesicles should preserve the integrity of donor and target compartment.
3. Fusion of vesicles with target membranes should be directed to ensure delivery only at the correct destination.

The composition of vesicles and target membranes allows for high specificity. First, vesicles exhibit a distinctive protein coat on their outside. The protein clathrin is the main component of so called **clathrin-coated vesicles**. This type of vesicle is best-studied and therefore, budding is well understood. The outer clathrin coat is connected via an adaptor protein called adaptin to specific receptors that span the lipid bilayer of the vesicle facing the inside to bind the cargo molecules, which display a **transport signal**. Once the cargo is bound by the receptor and the receptor is trapped by the adaptin that connects it with the clathrin coat, the vesicle starts to shape. When budding from the membrane of the donor compartment, a pit is formed, and finally, the vesicle is cut loose by a spiral-shaped protein called dynein. Pinching off the vesicle from the donor compartment requires energy as the lipid bilayer needs to be torn apart. The energy is provided by GTP hydrolysis, a process catalysed by dynein, which is therefore a GTPase. Once the vesicle is released to the cytoplasm, the clathrin coat is removed to enable fusion with target compartments.

Docking is a precondition for delivery. Matching the vesicles to the correct target membrane is ensured by **Rab proteins** and **tethering factors**. Rab proteins are GTPases that are specific to the cytoplasmic surface of the vesicle. Unique sets of Rab proteins are recognized by corresponding tethering factors. If there is a correct match between vesicle and target membrane, the Rab proteins are captured by the tethering factor. Additional strength in this interaction is provided by transmembrane proteins called **SNAREs**. A v-SNARE of the vesicle winds up with a t-SNARE from the target membrane thereby acting like a winch, pulling the vesicle close to the target membrane, enabling fusion of the lipid bilayers. Upon membrane fusion, cargo proteins inside the vesicles are released into the lumen of the target compartment.

5.3.1   Glycosylation of Proteins and Quality Control

Prior to onward transport, most of the proteins are chemically modified in the ER. Contrary to the cytoplasm, the environment in the ER is reducing, which results in the formation of disulfide bonds between reduced thiol groups from cysteine side chains. These bonds stabilize the protein before it is incorporated into the plasma membrane or secreted into the extracellular space. Rearrangements in S-S bonds are catalysed protein disulfide isomerases (PDIs). Another chemical modification of proteins in the ER is the attachment of sugar molecules. The process of adding sugars to proteins is called **glycosylation** and serves several purposes:

- Increase of protein stability by protection from proteolysis.
- Change of solubility and affinity as oligosaccharides are hydrophilic.
- Quality control and labelling because chaperones can hold back proteins.

To fulfil these diverse functions, glycosylation patterns must exhibit a remarkable diversity, which in fact they do despite maturing from a single precursor. Sugars are not added as monosaccharides one by one but rather a preformed oligosaccharide consisting of 14 sugar molecules that is covalently linked to the protein at once. Dolichol is a membrane lipid that faces the lumen of the ER with two phosphate groups and provides the sugar precursor. The oligosaccharide is transferred to the amino group of an asparagine within the polypeptide chain as it enters the ER through a translocator protein. The sugar transfer from dolichol to asparagine is catalysed by an oligosaccharyltransferase (OST), which is only active in the lumen of the ER. In subsequent reactions, the sugar precursor of the glycosylated protein is further processed in the ER and the Golgi apparatus.

Proteins are not shipped onward from the ER unless folded correctly. There are different chaperone molecules that steer correct folding of proteins in the ER. If folding however fails, proteins are exported back to cytoplasm in a process known as ER-associated protein degradation (ERAD). In case incorrectly folded proteins still accumulate in the ER, the **unfolded protein response (UPR)** is triggered. Accumulating protein is recognized in the ER by transmembrane protein kinases that facilitate the expression of additional chaperone genes while decelerating translation of other proteins by trapping ribosomes for the co-translational import of chaperones into the ER. As soon as proteins are folded correctly, and as long as they do not carry an ER retention signal, they are packaged into vesicles and exit the ER towards the Golgi apparatus.

The Golgi apparatus is a multi-layered structure with a polar morphology consisting of several flattened sacks called "cisternae". The side of the Golgi apparatus facing the ER is called *cis*-Golgi apparatus, opposed to the part facing the plasma membrane, which is called *trans*-Golgi network. Vesicles fusing to the *cis*-Golgi apparatus are originating from the ER. Vesicular transport from the ER to the *cis*-Golgi apparatus is referred to as anterograde trafficking. Cargo is immediately sent back if proteins contain the ER retention signal. This reverse route is referred to as retrograde trafficking. Within the Golgi apparatus, proteins are sorted according to their glycosylation patterns that are further processed, while proteins are migrated through the cisternae from the *cis*-Golgi apparatus to the *trans*-Golgi network. The sugar molecules linked to the proteins serve as a barcode that directs them to their destination. Vesicles budding from the *trans*-Golgi network are either directed through the early and late endosome to the lysosome for digestion, or they are sent within secretory vesicles to the plasma membrane. Vesicles with transmembrane proteins

Fig. 5.4 Vesicles are transported from the ER to the Golgi apparatus (anterograde trafficking) and back (retrograde trafficking). At the Golgi apparatus, secretory vesicles are budding that are delivered to the plasma membrane and cell exterior. Vesicles are also transported from the Golgi apparatus to the early endosome. The late endosome develops from the early endosome, which is exchanging material with the plasma membrane. From the late endosome, cargo is eventually transported to the lysosome for degradation

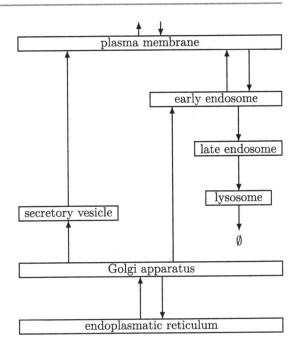

can be incorporated into the plasma membrane. Cargo within secretory vesicles or from vesicles of the early endosome can be released into the extracellular space. Withdrawal of receptors from the plasma membrane or uptake of extracellular material occurs by inward budding of vesicles from the plasma membrane. Early endosomes are created that mature into late endosomes, which transport cargo to lysosomes for digestion. Transportation routes of vesicles form a directed network with several intermediate stops (Fig. 5.4).

5.3.2 Exo- and Endocytosis

Along the major routes of vesicular transportation, two main directions can be discriminated:

1. **Exocytosis**: cytoplasm → Golgi apparatus → plasma membrane and
2. **Endocytosis**: plasma membrane → endosome → lysosome

For transmembrane proteins to be incorporated into the plasma membrane or proteins to be released into the extracellular space, the secretory pathway is followed in the process of exocytosis. The address label that directs proteins to the plasma membrane is a diacidic motif: DXD or DXE, where D stands for aspartic acid, E stands for glutamic acid and X can be any amino acid but proline. This motif is recognized in the Golgi apparatus and wrapped in secretory vesicles that fuse with the plasma membrane. There is a side-path from the *trans*-Golgi network to the cell periphery.

Proteins that are glycosylated with mannose 6-phosphate are shipped through early and late endosomes to the lysosome. This compartment is thus provided with many different hydrolysing enzymes for the digestion of proteins, nucleic acids, sugars and lipids. Thereby building blocks for cellular components (see Sect. 1.1) are recycled into the cytoplasm.

The lysosome is loaded with material for digestion and recycling from vesicles that are budding from the plasma membrane in the process of endocytosis. During vesicle formation, all membrane components are non-specifically taken up in the lipid bilayer of developing early endosomes. Maturation of late endosomes involves activity of ATP-driven H^+ pumps to maintain low pH within the endosome compartment. The acidic environment leads to the release of cargo from membrane-bound receptors, which can therefore be recycled directly back to the plasma membrane or degraded in the lysosome. Eventual uptake of membrane components by endocytosis needs to be balanced with membrane fusion of secretory vesicles to maintain the overall surface area of the plasma membrane constant.

A prominent example of receptor-mediated endocytosis is the **uptake of cholesterol**, an important component of cellular membranes. Cholesterol is secreted from the liver into the blood stream. Cholesterol is surrounded by particles of low-density lipoprotein (LDL) because it is water-insoluble. LDL is recognized by specific receptors on the cell surface and thereupon taken up into clathrin-coated vesicles. Due to the increased acidity in the endosomes, LDL particles are released from the receptors. In the lysosome, LDL particles are broken down by hydrolysing enzymes and the cholesterol molecules are freed into the cytoplasm for new membrane synthesis.

The cell's ability to regulate vesicular fluxes between the compartments is reflected by the three vesicle types that can be discriminated by their coat proteins. Besides clathrin-coated vesicles, there are two other types of vesicles: COP-I vesicles and COP-II vesicles. COP stands for coat protein. These three types of vesicles differ in their donor compartments, from which they are budding, and their target compartments, to which they are fusing (Table 5.2).

Table 5.2 Vesicle types, their donor and target compartments

Vesicle type	Donor compartment	Target compartment
Clathrin	*trans*-Golgi network	Late endosome
Clathrin	Plasma membrane	Early endosome
Clathrin	Early endosome	Late endosome
Clathrin	Secretory vesicle	*trans*-Golgi network
COP-I	*cis*-Golgi apparatus	*trans*-Golgi network
COP-I	*trans*-Golgi network	Plasma membrane
COP-I	*cis*-Golgi apparatus	Endoplasmatic reticulum
COP-II	Endoplasmatic reticulum	*cis*-Golgi apparatus

Procedure

Several technologies exist to identify and follow transport processes of proteins.

- There are **conditional mutant strains of yeast**, in which some proteins become inactive upon a temperature shift. As these mutant proteins are involved in transportation, respective processes are stalled once the temperature changes. For instance, proteins would accumulate in the ER, if the mutant were playing a role in COP-II mediated shuttling between ER and *cis*-Golgi apparatus.
- Proteins in certain organelles can be detected after **radioactive labelling**. Therefore, the polypeptides are translated in a cell-free system *in vitro* using radioactive amino acids. The labelled proteins can subsequently be incubated with isolated organelles. Co-sedimentation with these organelles in a centrifugation gradient implies that labelled proteins are indeed imported into the respective organelle.
- In contrast to the artificial cell-free system with protein synthesis from radioactive amino acids, the **green fluorescent protein (GFP)** allows to trace transportation processes in living cells by live-cell microscopy. GFP can be tagged by means of genetic engineering to potential cargo. As GFP is rather small, it hardly interferes with the function of the tagged protein [104]. GFP "glows" in green upon excitation with UV light, thus tagged proteins can be followed along the transportation routes within the cell.

◄

Summary

- Cellular membranes consist of a lipid bilayer, and the more long and saturated lipid tails they contain, and the more cholesterol is inserted, the more rigid are these membranes.
- Proteins are shuttled between nucleus and cytoplasm through nuclear pore complexes.
- Vesicular transport is the major way of intracellular trafficking, exocytosis and endocytosis.

Signalling

6

Contents

Introduction

Signalling is a non-biological paradigm for information processing. Biologists borrowed concepts from foreign disciplines, e.g. mechanical engineering, to understand the signal transduction of living systems. Conceptually, all cells communicate, in other words: they exchange information. Such processes exist on many different levels in space and time [61] and exhibit stunning parallels to information processing in technical devices (Table 6.1). A good source for spatial and temporal scales in biology is http://bionumbers.org. On the intercellular level, information is sent, received and processed between cells (Fig. 6.1). This concept applies, for instance, when the hormone insulin is released to inform other cells to increase sugar uptake. Likewise, all cells in our body serve as information processing units, which could be described as living computers (http://stanford.io/1qGnvrn).

The conversion from the input signal (S) to the output response (R) in biology is most commonly classified as either linear, hyperbolic or s-shaped functions (Fig. 6.2). The functional relation between S and R depends on the type of biological data and the measurement technique. Linear correlation is expected for protein abundance vs. messenger ribonucleic acid (mRNA) abundance in the physiological range because more mRNA should result in more protein. However, there is a tight and nonlinear interconnection between mRNA and protein regulation (see Chap. 4). Hyperbolic curves resemble enzymatic processes (see Sect. 4.3.1). A biological processor with a steep input-output conversion acts as a switch. If a switch-like mechanism is

© Springer-Verlag GmbH Germany, part of Springer Nature 2022
L. Adlung, *Cell and Molecular Biology for Non-Biologists*,
https://doi.org/10.1007/978-3-662-65357-9_6

Table 6.1 Networks deal with information. Parallels from technology to biology can be found at various levels

Technology	Biology
Internet	Population
Computer	Organism
Hardware	Tissues
Microprocessors	Cells
Logic section	Compartments
Transistors	Molecules

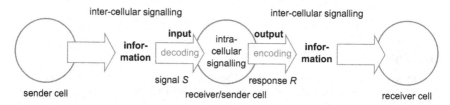

Fig. 6.1 Inter- and intracellular signalling. A sender cell sends information that are decoded as an input signal S by receiver cells. Intracellular signalling events process the input and encode an output response R that is sent as information to another receiver cell

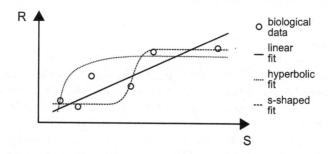

Fig. 6.2 Functional relations between the input signal S and output response R can be linear, hyperbolic or s-shaped, given experimental data (circles). Best fit depends on the biological context

investigated (see Sect. 6.5), or the detection of the output is insensitive to low S and saturates for high S, s-shaped curves will be recorded.

6.1 Molecular Switches between On- and Off-State

Apart from the information processing behaviour of entire cells, biological switches can be found on the molecular level. Inside cells, switches are molecules that can be toggled between the on-state (i.e. active) and the off-state (i.e. inactive) by binding or unbinding other molecules. Binding and unbinding can both, activate or inactivate, depending on the binding partners. The phosphate molecule (PO_4^{3-}), for example,

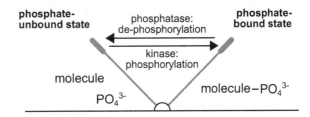

Fig. 6.3 Molecular switch depicted as a lever between a molecular phosphate (PO_4^{3-})-*un*bound state (left) and phosphate-bound state (right). Converting enzymes for de-/phosphorylation are called phosphatases/kinases

triggers the switching process. Enzymes that tack a phosphate onto another molecule are called kinases, whereas enzymes that pluck off the phosphate are called phosphatases (Fig. 6.3). An example for activation through phosphorylation (i.e. addition of PO_4^{3-}) is the transcription factor signal transducer and activator of transcription 5 (STAT5) (see Sect. 6.4) with one of its kinases Janus kinase 2 (JAK2) (i.e. activator) and phosphatases (i.e. de-activators) in the nucleus.

Guanosine triphosphate (GTP) and GDP work similarly to phosphate. Molecules can become active or inactive by being associated with GTP or GDP. G-protein coupled receptors serve as a paradigm (see Sect. 6.3.2). The conversion from GTP to GDP by GTPases (see Sect. 4.3.1; e.g. Ras as link to the Raf-MEK-ERK cascade, see Sect. 6.4.2) is called hydrolysis (as a water molecule is incorporated and a phosphate group is released). The reverse reaction is realized by an exchange of GDP with a new molecule of GTP. The catalysing enzyme for the latter substitution is classified as a guanine nucleotide exchange factor (GEF) like for the shuttling of molecules between cytoplasm and nucleus (see Sect. 5.2.1).

These regulatory principles are common throughout the entire signal transduction network, ranging from the perception of the input signal to processing events and the final output response (Fig. 6.4).

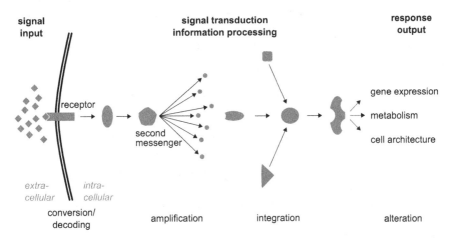

Fig. 6.4 Overview of signal transduction. Extracellular molecules can be sensed by receptors that convert the information input into an intracellular cue. Second messengers amplify the relayed signal and multiple signalling branches can be integrated to finally alter cellular structures and processes in response. Adapted from [6]

6.2   Molecules and Physicochemical Properties as Input from the Sender

The first layer of signal processing in biology is the information input. Diverse stimuli can be classified.

- Signalling molecules can serve as mediators of communication between cells (i.e. intercellular), such as hormones, growth factors and cytokines. Prominent examples are insulin and erythropoietin (Epo).
- Metabolic compounds play a role, too. Glucose, amino acids or adenosine triphosphate (ATP) are a valid readout for conditions within cells (i.e. intracellular).
- Chemicals (acids $[H^+]$, bases $[OH^-]$, metal ions $[Fe^{2+}/Fe^{3+}]$ and radicals $[\cdot O_2^-]$) characterize the cellular environment. Spatial surroundings could be acidic (cells in sour medium), or hypoxic (low-oxygen stress mediated through O_2 radicals).
- Physical properties such as light, temperature or pressure can also serve as signalling cues (e.g. heat shock in bacteria).

An example of mechanical confinements as input that influences emergent properties (see Sect. 6.6) is the physical pressure that seems to be the reason for altered (a)symmetry of cell divisions in mouse cells [103], though the precise mechanism remains to be resolved.

6.3   Receiving Cues via Cellular Receptors

The way incoming information is processed by a cell crucially depends on the layer of perception. Input is sensed by molecules called receptors that usually convert the signal from the outside of the cell (i.e. extracellular) to the inner, cytoplasmic part (i.e. intracellular). Some receptors however sense their ligands and trigger intracellular signalling from the cytoplasm only. Input molecules recognized by the receptor are called ligands.

There are three main types of receptors that can be discriminated by their structure and thus their functions. In the following, the three types will be introduced separately.

6.3.1   Ion Channels Maintain the Flow of Ions and Thus Information

Ion channels pass ions through the membrane with a simple pore-like structure. The driving force is the concentration gradient of the ions across the membrane. Ions are charged and thus create an electric potential. There is no need for an active transport. Channels can always be open, with ions diffusing passively through, or the pore opening and closing can be regulated.

An example is the sensation of salty (i.e. sodium chloride NaCl) taste. Our tongue is full of taste buds that contain taste receptor cells with so-called epithelial sodium ion (Na^+) channels. Those channels are always open and sodium ions from salty food diffuse through the pores into the cell to level out the concentration gradient (extracellular: high Na^+ ↔ intracellular: low Na^+). The sodium influx induces the opening of calcium ion (Ca^{2+}) channels (taste signal transduction reviewed in [19, 69]). Subsequent diffusion of Ca^{2+} into the cells leads to membrane depolarization (i.e. inversion of the electric potential across the cell membrane). This electric signal (i.e. action potential) can then be transmitted to the brain, which decodes the taste from the activation pattern of the taste buds' cells.

6.3.2   G-Protein Coupled Receptors Enable Switching Mechanisms

G-protein coupled receptors are capable of activating effector domains upon stimulation. Besides the extracellular part, which recognizes the ligand, they always contain seven transmembrane domains that penetrate the cellular envelope. On the cytoplasmic side of the membrane, they are associated with a GTP-binding protein (i.e. the "G-protein"). The activation process of G-protein coupled receptors can be summarized as follows:

ligand binding to the receptor
↓
conformational change in the structure of the receptor
↓
G-protein binding to the receptor
↓
GDP-GTP exchange (i.e. G-protein activation)
↓
subsequent target activation by the G-protein

The exchange between GDP and GTP provides a molecular trigger for intracellular signalling (see Sect. 6.1).

6.3.3   Enzyme-Coupled Receptors Regulate Complex Processes

Enzyme-coupled receptors either exhibit their own enzymatic activity or they are situated adjacent to enzymes. The purpose is to integrate different signalling inputs, which is strengthened by the fact that

$$\#(\text{receptor types}) \gg \#(\text{ligands}). \tag{6.1}$$

The combinatorial complexity and structural diversity of receptor complexes allows enzyme-coupled receptors to fulfil a myriad of functions. They are very sensitive and can be activated even by concentrations of growth factors in the pico-molar range (i.e. a single growth-factor molecule within more than a thousand trillion water molecules). The receptors are most commonly coupled to kinases, e.g. JAK2 for the erythropoietin (see Sect. 6.2) receptor (EpoR), which exhibits a K_D (i.e. dissociation constant) for Epo of $1.64 \cdot 10^{-10}\,\mathrm{M}$ [12]. This means, in equilibrium, the molar ratio of the concentration of unbound ligand [Epo] and free receptor [EpoR] to the concentration of ligand-receptor complex [Epo-EpoR] is K_D:

$$K_D = \frac{[\mathrm{Epo}] \cdot [\mathrm{EpoR}]}{[\mathrm{Epo\text{-}EpoR}]}.$$ (6.2)

If an enzyme-coupled receptor possesses intrinsic enzymatic activity, it is most commonly a tyrosine kinase. This means they are capable of transferring a phosphate group to tyrosine residues of other proteins (see Fig. 6.3). Tyrosine kinase receptors form dimers of single membrane-spanning domains, which phosphorylate each other. Phosphorylated tyrosine residues on the intracellular tails of the receptors serve as anchor points for adaptor molecules that relay the signal.

6.4   Signal Conversion in Protein Networks

The topological feature that diverse extracellular cues are integrated by a rather small set of core components is referred to as bow-tie structure [58] (Fig. 6.5).

Intracellular molecules are capable of amplifying, integrating or distributing the signal (Fig. 6.4). Three different transduction features can be classified, a combination of which enables regulatory complexity.

6.4.1   Second Messengers to Spread Information Rapidly

As the name implies, second messengers are a class of molecules that forms the next instance of signal transduction beyond the receptor. Second messengers are mainly induced via G-protein coupled receptors (see Sect. 6.3.2). Membrane-bound components are recruited and small molecules are subsequently released and diffuse quickly through the cytoplasm to spread the information. Some components serve as paradigms that should be known, e.g.:

- phosphoinositide 3-kinase (PI3K), producing
    - phosphatidylinositol *3,4 bis*phoshpate (PIP$_2$) and

Fig. 6.5 Bow-tie structure of signal transduction components. A heterogeneous ensemble of receptors feeds into a set of four core branches that divert into a multitude of various effectors and transcription factors

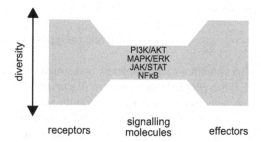

 – phosphatidylinositol *3,4,5 tris*phoshpate (PIP$_3$),
serves as a model system to understand regulatory features of tumour suppression (reviewed in [23]). Or
• phospholipase C, producing
 – diacylglycerol and
 – inositol 3,4,5 trisphoshpate (InsP3),
which is in turn responsible for Ca^{2+} waves [64] that are visualized particularly for fertilization http://bit.ly/1s22ycg (see Sect. 7.1).

A paradigm for a soluble second messenger that is not bound to the membrane but diffuses within the cytoplasm is the adenylate cyclase, which produces cyclic adenosine monophosphate (cAMP) that has been implicated, for instance, in nutrient sensing of yeast [101].

6.4.2   Kinase Cascades for Signal Relay

Connecting multiple kinases in a linear chain offers additional layers of interference for fine-tuning. Consecutive phosphorylation is sometimes facilitated by a scaffold protein, which keeps the kinases in spatial proximity. Each kinase can be regulated by incoming (e.g. negative feedbacks, see Sect. 6.5) and outgoing (i.e. additional target substrates) connections. In a prominent case, the hierarchy of kinases within the cascade can be inferred by the name: The extracellular signal-regulated kinase (ERK) is a mitogen-activated protein kinase (MAPK), while its kinase, mitogen/extracellular signal-regulated kinase (MEK), is a MAPK kinase (MAPKK), and the kinase of MEK is Raf, a MAPKK kinase (MAPKKK).

This kinase cascade (less Ks mean substrate) is involved in proliferation of blood progenitor cells in mice [94].

6.4.3   Transcription Factors Control Gene Expression

Downstream to receptors and possible second messengers in the signal transduction hierarchy, there are proteins modulating the initiation of transcription (see Chap. 4) that are referred to as transcription factors. They can represent a direct link from the receptor to target-gene expression. Upon signalling, transcription factors are shuttled to the nucleus and bind to the DNA. The order of signalling events is as immediate as follows:

receptor activation
↓
transcription factor activation
↓
shuttling from the cytoplasm to the nucleus
↓
target-gene control

Such a route is direct and thus highly efficient and specific without intermediate branching steps. The EpoR-JAK2-STAT5 axis serves as a paradigm in this respect. Despite the immediate link from the receptor EpoR with its adjacent kinase JAK2 to the transcription factor STAT5, there is the possibility for regulatory mechanisms. STAT5 is activated through phosphorylation (see Fig. 6.3), and two activated molecules form a complex. These dimer structures exhibit distinctive transport and DNA-binding kinetics as a base for mechanistic diversity [16].

6.5   Characteristic Motifs and Behaviours

The elements of signal transduction that were introduced above are highly interconnected in molecular networks. The wiring (i.e. the network structure) encodes distinctive functions in motifs and sub-modules. Pathways must not be thought of as linear chains. There are many interconnections. Feedback loops are either positive or negative with respect to the conversion of the input signal S to the response R (Fig. 6.6). Positive feedback loops further increase R with increasing S, while R is reduced in negative feedback loops by rising S.

These loops render molecular pathways non-linear. The shape of signal-response curves depends on the topology of the underlying network motif. The structure of the sub-modules and their respective function are intimately related and gave name to exemplary "sniffers", "buzzers" and "toggles" [106].

A sniffer for example is a motif that responds transiently to signal input, but the steady state of the response R is independent of the signal S. The sniffer is named after the molecular pathway responsible for the perception of odour within the sensory neuronal cells of our nose. Its structure (Fig. 6.7A) entails a negative feedback loop because the component X that is induced by the signal input S diminishes the response R. More S results in less R, which indicates a negative feedback. In simple terms: For high input, the systems show a transiently high response. But high S results in high X and a correspondingly strong reduction of R, which thus always returns to the same steady state (Fig. 6.7B). This return is called "adaptation". The system adapts to the input signal and its response returns to basal activity. This feature can be further exploited as the model files are provided as online material to be simulated with the freely available D2D Software.

Fig. 6.6 Process diagrams of feedback loops. (**A**) Positive feedback loop further increases response R with more signal S. Production of R is additionally catalysed by intermediate species X. (**B**) Negative feedback either via inhibition of synthesis of R through X (left motif) or degradation of R catalysed by X (right motif). Visualisation adopted from [59]

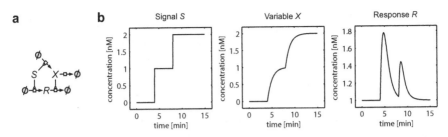

Fig. 6.7 Sniffer network motif and dynamic behaviour. **A**) Signal input S induces production of response R and production of intermediate species X, which catalyses the degradation of R. **B**) Temporal evolution of the three components S, X and R. With more S, R is transiently increased but reset by X to its basal steady-state level. Adopted from [106]

Thus, our sensory system is always reset to its basal minimum state and we adapt even to strong smells. However, excitation (i.e. a response) occurs for every new input smell. Similarly, bacteria sense nutrients in their surrounding. Adaptation is an evolutionarily conserved property. Bacteria adapt to chemical gradients but can respond to every new input via molecular feedback loops [1]. Thereby they *robustly* sense nutrients and swim towards the attractant. Robustness in this context is defined as a small change in the variable X for small changes in the parameter p:

$$\frac{\delta X}{\delta p}. \tag{6.3}$$

If the system's parameter p underlies minor fluctuations, be it through some input perturbation, the readout variable X remains unchanged. In contrast to robustness, sensitivity means that the system responds to subtle changes in p with a large change in X (Eq. 6.3). In summary:

- a system is robust if its output is not altered despite perturbation and
- a system is sensitive if small changes can be sensed and alter the output.

6.5.1 Signal-Response Curves

The term sensitivity is also used in the context of drug research. If the effect of a newly developed drug is investigated, the response (of the cellular system, the animal or the patient) to different *doses* of the substance is evaluated. That is why the signal-response curve is referred to as a *dose*-response curve and sensitivity can be depicted in such a curve.

A measure for sensitivity is the dose at which a half-maximal response is observed (Fig. 6.8). This parameter is denoted as $EC50$ (i.e. half-maximal effective concentration) [27] if the response increases with dose. In case an inhibitor is tested and the response is thus diminished with increasing doses of the drug, the parameter is called $IC50$ (i.e. half-maximal inhibitory concentration). Anyway, it signifies the

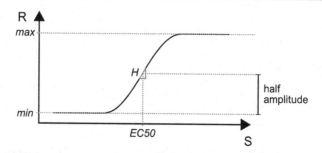

Fig. 6.8 Dose-response curve. Response R changes with input S between minimum *min* and maximum *max* with slope H and a half-maximal effective concentration at $S = EC50$

dose at which half the response amplitude between minimum *min* and maximum *max* is reached. A measure for the steepness of the curve is the Hill coefficient H that indicates the level of cooperative effects of the drug. Cooperative binding would mean that binding of a single ligand to the receptor facilitates binding of another one. A formula for a four-parameter Hill equation with those parameters is

$$R = max \cdot \frac{min - max}{1 + \left(\frac{S}{EC50}\right)^{-H}}. \tag{6.4}$$

The parameters can be estimated given experimental data and the dose-response behaviour can be quantified. The sensitivity towards the drug indicates susceptibility of the system towards the substance as well as the efficiency of the drug.

6.5.2 Time-Course Curves

Dose-response behaviour is no static property, it underlies a dynamic. The drug needs to be taken up by the body (pharmaco*kinetics*) and cells respond with time (pharmaco*dynamics*).[1] Sensitivity (see Fig. 6.8) can be reflected in temporal profiles of signal transduction, too. If one parameter fluctuates, the systemic variables might also change. Furthermore, all biochemical processes are stochastic (Brownian motion of particles that diffuse through a cell) and exhibit a certain degree of noise. Sources of noise can be:

- intrinsic noise: fluctuations in cellular protein concentrations,
- extrinsic noise: alterations in kinetic rates of cellular processes and
- measurement noise: errors and imprecision during experimental proceedings and data acquisition.

---

[1] Pharmacokinetics is what your body does to the drug. Pharmacodynamics is what the drug does to your body.

Signalling events are commonly investigated in time-course experiments via a technique called immunoblot or western blot.

> **Immunoblot aka. Western Blot**
> Cells are lysed—the membrane (see Chap. 5) is chemically disrupted—so that their proteins become accessible. The cellular proteins are then separated by size and transferred to a membrane where they can be detected via antibodies (see Sect. 8.3.2). As readout serves the signal intensity for the detected protein that is scanned from the membrane. The bigger the spot (called a "band") of the respective protein on the membrane, the stronger the corresponding biological response.

In this exemplary case, mouse erythroid progenitor cells are stimulated with Epo and the corresponding phosphorylation (i.e. activation) of STAT5 with time is measured. Figure 6.9 shows the detected amounts of phosphorylated STAT5 (pSTAT5) at the given time points. The bigger the spot (called a "band"), the more pSTAT5 is present (Fig. 6.9). As the transcription factor is shuttled to the nucleus during its course of activation (see Sect. 6.4.3), STAT5 was analysed in the cytoplasm and the whole-cell lysate (i.e. cytoplasm + nucleus).

Once the measurement is quantified, the time-course curve can be plotted and characterized (Fig. 6.10). The amount of the protein of interest at the beginning of the experiment at $t = 0 = t_0$ is referred to as its initial concentration or basal level. Local maxima that are reached in the course of the experiment are called "peaks" and the time until they are reached is the respective peak time. Afterwards, the system usually adapts to a steady-state level. Dynamics can be either transient, if the steady state is low (Fig. 6.10), or sustained, if the steady state remains high after peak activation.

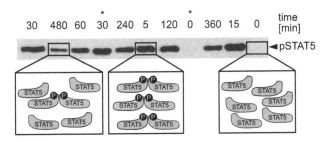

Fig. 6.9 Western blot data of phosphorylated STAT5 (pSTAT5) in whole-cell lysate (*) or cytoplasm (otherwise). Erythroid progenitor cells from mice were stimulated with 5 U/ml Epo for indicated times. Order of samples was randomized to reduce correlated errors [93]. The detected amount of pSTAT5 was quantified and plotted in Fig. 6.10. Data from [4]

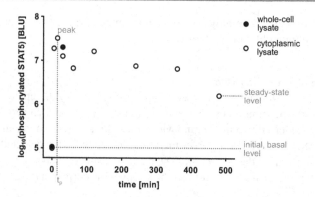

Fig. 6.10 Time-course experiment. Erythroid progenitor cells from mice were stimulated with 5 U/ml Epo. The stimulation was terminated either via whole-cell or cytoplasmic lysis. The phosphorylation of STAT5 was analysed via western blot (Fig. 6.9). After the peak time t_p of 15 min, signalling returns to a low steady state on a logarithmic scale. BLU = Boehringer Light Units

STAT5 activation is transient because it is reset by negative feedback regulators (see Sect. 6.5). Both cytokine-inducible SH2 domain-containing protein (CIS) and suppressor of cytokine signalling 3 (SOCS3) inhibit STAT5 activation and thus reduce the amount of pSTAT5 with time.

6.5.3 Complex Interactions

After the first peak, dynamics of biological systems are not always monotonous. Temporal responses $\frac{dR}{dt}$ could resemble periodic behaviour. Oscillations occur if certain preconditions are met [83]:

- feedback regulation (see Fig. 6.6),
- sufficient level of non-linearity and
- separated time scales.

Separated time scales mean that the reactions exhibit different velocities. Kinetics are balanced and occur with a sufficient delay to allow changes that are about to be levelled out. Due to dependence on the mentioned preconditions, oscillations occur only in a subregion of the parameter space. An example of molecular oscillations is the core signalling component (see Fig. 6.5) nuclear factor kappa-light-chain enhancer of activated B cells (NFκB) that is known to cycle periodically between cytoplasm and nucleus for the control of gene expression [77].

The interplay and interdependencies of molecular components become more complex as the network size grows. Pathways exhibit crosstalk and sophisticated response patterns are created [95]. Complexity is required because the outcome of signalling events are decisions to alter cellular properties in response to the multi-parametric environment, so the cell has to process and integrate several inputs (see Sect. 6.2) at a time.

6.6   Emergent Properties as Output

Cells integrate multiple signals to choose their fate in terms of:

- survival or programmed cell death,
- proliferation or quiescence,
- migration or residence,
- metabolic alterations and
- differentiation.

The term "emergent property" means literally the outcome of signalling events. Multiple properties have to be measured to assess the output upon information processing. To decide, whether a cell undergoes programmed cell death, morphology and molecular components need to be evaluated (see Sect. 9.2.1). In the exemplary case of the EpoR-JAK2-STAT5 axis, the signalling input Epo is connected to the output response of cell survival. Among the target genes of STAT5 are some factors that inhibit cell death. But how is the amount of Epo that is sensed by the cells converted to the alteration of gene expression? Fine regulation via the negative feedback regulators, CIS and SOCS3, allows erythroid progenitor cells in mouse to respond specifically to a wide range of Epo concentrations. By mathematical modelling, it was shown that the amount of activated STAT5 (i. e. pSTAT5) in the nucleus of the cells is a good predictor for the fraction of surviving cells in a population [10]. Cell morphology and different molecular components implicated in cell death were measured to assess the emergent property of survival as an output response of the erythroid progenitor cells to the input signal Epo.

6.7   Signal Input, Processing and Output Happens on Various Levels

Cellular signalling can be diverse as many different components are involved and interconnected. As introduced in Fig. 6.5, there are a few central components that should be known as well as some core features of molecular networks (Fig. 6.6). The overview given in Fig. 6.4 is completed by names for molecules that serve as paradigms in Table 6.2, most of which were introduced in this chapter.

Table 6.2 Paradigms for different signalling pathway components. *IN* = input, *REC* = receptor, *KIN* = kinase, *2nd* = second messenger, *INT* = integrator, *TF* = transcription factor, *TG* = target, *OUT* = output, *REF* = reference

IN	REC	KIN	2nd	INT	TF	TG	OUT	REF
Epo	EpoR	JAK2	–	–	STAT5	CIS, SOCS3	Survival	[10]
		PI3K	AKT	–		Cyclin D2, Cyclin G2	Cell cycle	[3]
		–	ERK	–	–	–	Proliferation	[94]

Summary

- Receptors sense information input into the cell by binding cognate ligand molecules and triggering intracellular reactions.
- Information are being processed inside the cell through signal transduction pathways.
- Inhibition and activation can be intertwined in feedback loops, which render cellular output highly non-linear in space, time and dose-response.

Cell Cycle

7

Contents

Introduction

The basis for reproduction of all organisms is a process called the cell cycle [75]. As the name implies, the cell cycle is an iterative, periodic chain of events, in which the cell undergoes repetitive cycles of growth and division. Proper timing and execution of the cell-cycle programme is vital for the development and maintenance (i.e. homeostasis) of our tissues and organs. Malfunctioning of the cell cycle is connected to uncontrolled growth and tumour formation. The cell cycle is therefore tightly regulated by cellular signal transduction networks (see Chap. 6). Several checkpoints exist to ensure that the cell cycle only progresses if all processes are performed correctly. Unless the preconditions are fulfilled (e.g. doubling of the DNA before cell division), the process is halted.

In principle, the cell cycle can be depicted as a circle (Fig. 7.1) with two clearly defined events, namely DNA synthesis (S-phase) and mitosis (M-phase). In between those two discrete events, the cell prepares for DNA doubling and division in so-called gap (G)-phases. Throughout the cell cycle, the cell replicates its contents e.g. membranes, cytoplasmic organelles, structural proteins, rRNA and mRNA. This continuous process is difficult to observe compared to doubling of the chromosome set during S-phase and the distribution of cellular mass to an upcoming pair of daughter cells during M-phase. Thus the name G-phase is a bit misleading. In fact, cells actively grow in the G_1 phase if they are not interrupted at a quality control checkpoint.

© Springer-Verlag GmbH Germany, part of Springer Nature 2022
L. Adlung, *Cell and Molecular Biology for Non-Biologists*,
https://doi.org/10.1007/978-3-662-65357-9_7

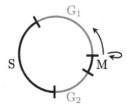

Fig. 7.1 Overview of cell cycle. Two defined phases of mitosis (M) and DNA synthesis (S) are interrupted by two gap phases (G_1 and G_2), in which the cell grows and prepares (during G_1) for replication of DNA, and (during G_2) for cell division (↻). Length of cell-cycle phases adapted for a 20 h cell cycle of HeLa cells from [42]

Besides accuracy, the correct timing of the cell cycle is essential. The length of the individual phases needs to be well coordinated. Cells dividing too rapidly can develop malignancies while slowly dividing cells can impair tissue regeneration following lesions. It is very important to maintain the balance for fast and precise completion of the cell cycle. What are the components that constitute the underlying molecular machinery? The following section will shed light on the discovery of core cell cycle components and their mode of action.

> **Conventions**
> Non-dividing cells are sometimes supposed to be arrested (i.e. stopped) in a phase called G_0 prior to entering cell cycle.
>   For convenience all phases except the M-phase are often summarized as Interphase $= G_1 + S + G_2$.

7.1   Cell-Cycle Model Systems

Many important observations regarding the cell cycle have been made in model organisms such as yeast, fruit fly, nematode worm and frog. Table 7.1 gives an overview of some species predominantly used in cell-cycle research.

Yeast Their genomes are simple, their doubling times are short and their tradition as research subjects is long-standing, even though the two unicellular fungi, budding yeast and fission yeast, exhibit slightly different mechanisms of division. Haploid cells can be maintained in culture and genetically manipulated. Conditional mutants were used to identify essential genes for cell-cycle progression. Genes were non-functional under certain conditions, e.g. low temperature. These conditional mutants were arrested in cell cycle when shifted to low temperature. Otherwise, the wild-type cell-cycle progression was observed. Thus, factors were identified that are respon-

Table 7.1 Exemplifying species used in cell-cycle research and main advantages for their use

Organism	Species	Advantage
Budding yeast	*Saccharomyces cerevisiae*	Simple Eukaryote and straightforward genetic analysis
Fission yeast	*Schizosaccharomyces pombe*	Simple Eukaryote and straightforward genetic analysis
Fruit fly	*Drosophila melanogaster*	Cell cycle control mechanisms with intermediate complexity
Nematode worm	*Caenorhabditis elegans*	Well studied for mitosis and division
Frog	*Xenopus laevis*	Giant unfertilized eggs, maternal proteins in embryo

sible for chromosome duplication and segregation as yeast cells undergo repetitive cycles of growth and division.

Fly The genome of the fruit fly contains $\approx 14,000$ genes, twice as many as yeast but only half the number of the human genome. Its chromosome set is diploid, which makes genetic manipulation a bit more intricate than for haploid yeast. Changes in cell cycles during the development of this multicellular organism make it nonetheless a suitable subject to study control mechanisms. During the first rounds of cell cycle after fertilization, only DNA and nuclei are doubled but neither cytoplasmic growth nor cleavage is observed. Such a cell with multiple nuclei is called syncytium. Nuclei locate at the outer rim of the embryo and are then encapsulated by membranes. This process of cellularization depicts an important step during early divisions in the fruit fly.

Frog The unfertilized egg spans a diameter of about 1 mm. It is possible to inject various substances into this single cell or extract small parts and study them in a test tube. This is how the maturation- or later mitosis-promoting factor (MPF) was found. Mock fertilization of these eggs with an electrical pulse in a Petri dish can trigger the release of calcium ions inside the cell (see Sect. 6.4.1).

Upon fusion of the haploid egg nucleus with the haploid sperm nucleus, the cells of the early embryo grow rapidly because they are provided with maternal RNA and protein as starting material.

7.1.1 Cellular Contexts to Study the Cell Cycle

Those model organisms are all suitable for the discovery of key, conserved mechanisms due to their low complexity and the available analysis tools. Nonetheless, the ultimate goal is the understanding of the mammalian cell cycle and finally intervention in the context of deregulation and human diseases.

Sources of human material are limited. Biopsies provide healthy or diseased tissue for patients and are thus very precious. As mice are genetically similar to humans

($\approx 85\%$ sequence identity of protein-coding genes)[1], they are used as model organisms for human disease such as type 1 diabetes [9]. Regulatory mechanisms of the genomes were diverging during the evolution of the two lineages (only $\approx 50\%$ of non-coding DNA regions are shared between mouse and human) [21]. However, newly established tools allow to study the function of certain genes in the living animal (i.e. *in vivo*).

A tissue sample (so-called primary material) taken out (i.e. *ex vivo*) of the living animal is only able to survive for a short time. The number of cell cycles and divisions is limited. The environment during cultivation of the primary cells in a plastic dish (i.e. *in vitro*) is too artificial. It does not resemble physiological conditions. Tumour cells survive and divide almost infinitely even *in vitro*. They possibly acquired growth-promoting mutations to become immortalized. Also, some non-tumour cells are transformed so that they can sustain with serum and sufficient amounts of growth factors *in vitro*. These cell lines can be expanded to generate enough biological material for cell-cycle studies. Notably, such cell lines may exhibit up-regulated cell-cycle-associated functions as they evolved towards rapid growth during cultivation [85].

7.1.2   Discovery of Core Cell-Cycle Components

The Nobel Prize in Physiology or Medicine 2001 was awarded jointly to Leland Hartwell, Tim Hunt and Paul Nurse for their discoveries of key regulatory components of the cell cycle. They described the mode of action of cyclins and cyclin-dependent kinases (CDKs) in regulation of DNA synthesis, chromosome separation and cell division.

Hartwell identified cell division cycle (*cdc*) genes in budding yeast (Table 7.1). He defined checkpoints for control of correct execution of preceding cell-cycle phases. Nurse worked with fission yeast and found the gene *cdc2* as a regulator for both, transition from G_1 to S-phase and transition from G_2 to M-phase. The homologous gene in budding yeast is *CDC28* and the corresponding human gene is *cdk1*, a CDK. Hunt discovered cyclins, the counterpart required for functioning CDKs. He radioactively labelled those proteins and found them to be degraded periodically at every M-phase in eggs of sea urchins.

While the concentrations of CDKs remain constant throughout the cell cycle, cyclins, as their name implies, are repeatedly synthesized and degraded (Fig. 7.2). CDKs depend in their activity on the presence of cyclins. Only if cyclins are bound to CDKs, they can phosphorylate their substrates. This central mechanism is conserved across species from yeast to insects, plants, animals and humans.

---

[1] genome.gov/10001343, last opened March 10, 2022.

Fig. 7.2 Concentrations of cyclins and CDKs during inter- and M-phase. While cyclins are constantly synthesized and degraded, concentrations of CDKs do not change, but their activity depends on the presence of cyclins

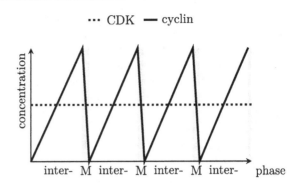

7.2   Checkpoints Regulating the Cell Cycle

In an ideal environment, unrestrained cell-cycle progression occurs. Cell cleavage (i.e. division) without delay happens for cells of the early embryo, which do not grow but rapidly progress from the 1-cell stage to the $2^6 = 64$-cell stage. Such accelerated divisions are also observed for cells (and unicellular organisms, e.g. yeast) in culture because they are usually bathed in growth factors, and growth is therefore not the rate-limiting step anymore [85]. Cell-cycle progression in tissues of higher organisms is otherwise restricted to ensure maintenance of size and to exert specific functions. The timing of the cell cycle is regulated by extracellular and intracellular signalling cues. Not only nutrients and growth factors but also physicochemical properties of the cellular environment compose regulatory factors from the outside of the cell, as we have learned before (see Sect. 6.2). Intracellular information consist of the signalling programme and gene expression status. For both, extracellular and intracellular regulation of cell-cycle progression, exist a multitude of involved components that differs in name and associated functions between species and organisms.

Conceptually, the wheel of the progressing cell cycle (Fig. 7.1) is spinning unless diverse braking mechanisms prevent the cells from cycling too fast. There is a restriction point in the G_1-phase of the cell cycle that is often displayed as a point of no return. It has been speculated that once all extracellular and intracellular requirements for passing the restriction point are fulfilled, the cell completes the current cycle even if growth factors are withdrawn. However, if the cell does not sense enough mitogens before passing the restriction point, it remains quiescent. Growth factors and mitogens are occasionally distinguished by their function. Growth factors promote growth and mitogens stimulate division. However, as we will see in Sect. 7.3, it is sometimes hard to discriminate those two processes.

7.2.1   The Restriction Point Mediated by Rb and E2F

The all-or-none decision between cell-cycle progression or cell-cycle stop is achieved by a switching mechanism (see Sect. 6.5) of the E2F protein [114]. E2F is a transcription factor for genes involved in DNA replication and thus cell cycle progression to S-phase. An important feature is the autocatalytic activity of E2F, which activates transcription of its own gene. Such a positive feedback loop characterizes the bistable switch between active and Rb-inactive E2F. Repression of E2F is mainly coordinated by the retinoblastoma (Rb) protein. Rb contains a binding pocket for E2F and is therefore a pocket protein. If E2F is bound by Rb, it cannot bind to DNA and is repressed in its function as a transcription factor. Phosphorylated Rb (pRb) releases E2F that can exert its function.

Rb as such is the substrate of CDK2 or CDK4/6. These kinases are only active when bound to their respective cyclins E or D. This redundancy of Rb-phosphorylating enzymes ensures robustness of the pathway against perturbations of individual components and enables integration of several signalling branches. Extracellular growth factors activate signalling pathways through MAPK and AKT (Table 6.2), which directly or indirectly lead to expression of the cyclin E and the stable assembly of a complex between cyclin E and CDK2. This process is inhibited the co-factors p21 and p27. The latter one represents also the internal state of the cell because it is degraded as the cell cycle starts. This degradation is initialized by mitogen-induced phosphorylation of p27, which is labelled for degradation by ubiquitination (see Sect. 4.3.1). The ubiquitination is catalysed by a generic Skp, Cullin, F-box containing (SCF) complex that specifically binds phosphorylated p27 by an adaptor protein called Skp2. The gene encoding for Skp2 is among the targets of E2F [116] creating another feedback loop in this regulation scheme of cell-cycle progression (Fig. 7.3).

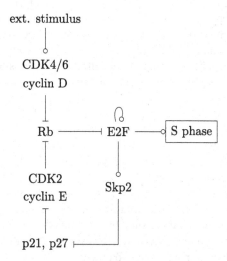

Fig. 7.3 External stimuli and intracellular signalling determine cell-cycle progression to S-phase. Growth factors and mitogens trigger complex formation between cyclin D and CDK4 or CDK6 that phosphorylate Rb. Phosphorylated Rb releases E2F that activates transcription of its own gene as well as S-phase promoting genes. Induction of Skp2 degrades p27, which otherwise would, together with p21 inhibit complex formation between cyclin E and CDK2, which also phosphorylates Rb. Inhibition: ⊣. Catalysis: ─o

7.2.2   The DNA-Damage Response

A safeguard mechanism exists to ensure correct duplication of the cellular DNA content, once the cell passed the restriction point and entered S-phase. Eventual DNA damage is sensed that triggers DNA repair and cell-cycle arrest. This response can be considered a DNA damage checkpoint.

Damaged DNA is initially bound by one of the two factors adenine-thymine mutated (ATM) or ATM- and Rad3-related (ATR), which are evolutionary conserved and named after the context of their discovery. ATM and ATR are both kinases that recruit other adaptors or mediators to the local site of DNA damage. Phosphorylation of those regulatory proteins results in the activation of effector kinases, namely checkpoint kinase (Chk)1 and Chk2. The mechanism of response and the involved factors depend on the type of DNA damage. Three different scenarios exist. A DNA damage response is triggered in severe cases of these three phenomena.

- **Block of the progression of the DNA replication fork.** A stalled replication fork can be induced by abnormal DNA structures and *vice versa*. If DNA replication does not proceed, M-phase entry is also blocked. Cell division prior to the completion of DNA duplication is referred to as mitotic catastrophe and should be prevented to avoid genomic instability.
- **Single-strand DNA damage.** Errors on one strand of the DNA can be corrected by base- or nucleotide-excision repair. The incorrect DNA stretch is cut out and the complementary strand provides the template for re-synthesizing the original sequence.
- **Double-strand breaks.** If both strands of the DNA do break, loose ends could lead to chromosome rearrangements. Non-homologous ends are joined in G_1-phase to prevent this. Disruption of protein-coding genes in humans is unlikely given that more than 98% of the human genome are non-coding sequences [32]. Double-strand breaks can be repaired by homologous recombination with sister chromatids after chromosome duplication upon S-phase completion.

ATR is recruited by the replication protein A (RPA) to the single DNA strands of the stalled replication fork, or upon base or nucleotide excision. RPA is a single-strand binding protein that stabilizes the DNA strand. Additionally, a sliding clamp (9-1-1 complex) and a clamp loader (Rad17-RFC) are recruited similarly to conventional DNA replication (see Sect. 2.3) for stabilization of the DNA structure. It is important to stabilize the stalled replication fork that would otherwise collapse and create more severe DNA damage. The inhibition of complexes between CDK2 and cyclin A blocks the initiation of DNA replication, while the replication fork is stalled. Inhibition of the phosphatase Cdc25 leaves the complex between CDK1 and cyclin B inactive, which would be required for M-phase entry. Recruitment of a DNA-unwinding helicase facilitates the resumption of DNA synthesis.

ATM is recruited by the Mre11, Rad50 and Nbs1 (MRN) complex to double-strand breaks. The MRN complex holds together the otherwise loose ends of the broken DNA helix. Histone modifications are mediated by ATM to alter local chromatin structure. Phosphorylated histones lead to an increase in the local concentration of cohesins that bring sister chromatids in closer proximity to allow for homologous recombination. If double-strand breaks occur in G_1-phase and DNA was not yet replicated, Chk phosphorylates the phosphatase Cdc25. Phosphorylated Cdc25 is targeted for degradation by ubiquitination through the SCF complex. Inhibitory phosphorylation at CDK2 remains and cells therefore do not pass the restriction point and arrest in G_1-phase.

Long-term DNA damage response involves p53, which induces expression of target genes responsible for cell-cycle arrest and programmed cell death (see Sect. 9.2.1). Regulation of p53 is threefold. First, p53 is phosphorylated by Chk and therefore stabilized. The nuclear export signal of phosphorylated p53 is blocked ensuring p53-mediated gene expression in the nucleus. Second, lysine residues of p53 are acetylated by p300, which prevents p53 from ubiquitination and degradation. Last, p53 is regulated by the ubiquitin ligase Mouse double minute 2 homolog (Mdm2) because it targets p53 for degradation and blocks the DNA-binding capacity of p53. Activity of Mdm2 was reported to be reduced upon DNA damage [20]. Among the p53 target genes is also p21, which inhibits the complex between CDK2 and cyclin D (Fig. 7.3). B cell lymphoma 2 (Bcl2)-family members, which initiate programmed cell death, are induced by p53 as last option to prevent detrimental effects of propagated genomic instability in case DNA repair did not succeed.

7.2.3   Hyperproliferative Stress

The activation of p53 leads to cell-cycle arrest. Tumours are characterized by uncontrolled growth and division, which can be suppressed by p53. Therefore, p53 is considered a tumour suppressor. On the other hand, genes encoding for factors responsible for proliferation are called oncogenes as they could induce cancer if they are overexpressed or hyperactive (see Sect. 9.2). Examples for oncogenes are the genes encoding for Myc and Ras. Both are stimulated by growth factors and activate E2F for cell-cycle progression. A negative feedback exists to prevent cells from uncontrolled proliferation in case of hyperactivation of these factors in a scenario called hyperproliferative stress. It could be required if cells are exposed to an excess of growth factors or oncogenes are mutated to become hyperactive or constitutively expressed.

Mdm2 is inhibited upon upregulation of Myc or Ras, which in turn releases p53 to induce expression of p21. As we have learned, p21 represses the complex between CDK2 and cyclin E, which is required for G_1/S transition (Fig. 7.3). This mechanism renders cells more likely to arrest in cell cycle or induce p53-mediated programmed cell death instead of growth and division without control.

7.3 Conceptual Views on Cell-Cycle Phenomena

It has been hypothesized that variability in cell cycle length mainly arises from different growth times in the G_1-phase [89]. Once cells are stimulated by growth factors, or committed to a developmental programme, they start to cycle. Cells that are bigger grow for longer before they have acquired sufficient biomass to proceed in the cell cycle, at least if growth and division in response to growth factor stimulation occur subsequently in a cell. Both processes could also be regulated in parallel or coordinated by diverging signalling pathways. These three scenarios (Fig. 7.4) are not mutually exclusive and could be combined depending on the cell type, the internal signalling state and the environmental condition.

A prerequisite for the understanding of the interdependence between cell growth and cell division is to know the individual processes as such.

7.3.1 Cell Growth as Biomass Acquisition

The size of a cell scales with its DNA content. We have seen in a simple paradigm (see Fig. 3.1): DNA $\rightarrow$ mRNA $\rightarrow$ protein; and protein contributes the most to dry mass of a cell. If a cell needs to grow, it needs to acquire biomass, and thus it needs to produce protein. For protein synthesis, translation turned out to be the rate-limiting step because the transcription rate for induced genes exceeds usually the translation of the respective mRNA molecules:

$$\text{DNA} \xrightarrow[k_1]{\text{transcription}} \text{mRNA} \xrightarrow[k_2]{\text{translation}} \text{protein},$$

with $k_1 \gg k_2$. The cell needs more ribosomes to boost translation and therefore increases the production of rRNA and ribosomal protein in response to nutrients and growth factors.

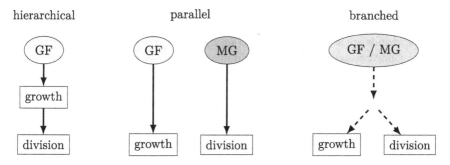

Fig. 7.4 Growth and division triggered by growth factors and/or mitogens can be coordinated in three different ways: hierarchical, in parallel or by branched signalling pathways. Mostly a combination of these mechanisms exists that is hard to disentangle. GF: growth factor, MG: mitogen. Adapted from [75]

The cell undertakes additional measures besides an increase in the number of ribosomes to grow and accumulate biomass. The entire metabolism is shifted. A central regulatory component of cellular metabolism in mammals is the mammalian target of rapamycin (mTOR). Its name is derived from the growth inhibitor rapamycin, which blocks mTOR activity. Uptake of amino acids and expression of metabolic enzymes is mediated by mTOR. Another downstream target activated by mTOR in response to growth factor stimulation is the ribosomal subunit S6. Translation is further enhanced through mTOR by phosphorylation of the eukaryotic initiation factor 4E (eIF-4E)-binding protein (4E-BP). Phosphorylated 4E-BP releases eIF-4E, which then promotes initiation of translation (see Sect. 3.3.1).

The cellular biomass could be even more increased by further stabilizing proteins thereby preventing their degradation. It has been observed that application of proteasome inhibitors reduces activation of downstream components of mTOR such as 4E-BP [60]. But whether mTOR regulates cell growth also by directly interfering with the proteasome is unclear.

7.3.2   Cell Division as Biomass Distribution

Accumulated biomass has to be distributed to daughter cells for division. This dynamic process during M-phase includes not only re-organization of proteins but also DNA. Mitosis represents the last phase of the cell cycle prior to cell division and is divided into five stages that can be morphologically discriminated.

1. **Prophase** is initiated by condensing chromosomes in the nucleus. The chromatin state is switched from the open euchromatin state, which still allows for DNA replication and transcription, to the tightly packed heterochromatin state, in which DNA is associated to histones and other proteins. The heterochromatin state in the prophase of mitosis makes chromosomes visible in their classical X shape.
2. **Prometaphase** starts with the breakdown of the nuclear envelope. Microtubules form a bipolar spindle apparatus and associate with kinetochores, which are located at the centres of the replicated sister chromatids (see Sect. 2.1). Sister chromatids are attached to opposite spindle poles. The spindle assembly checkpoint prevents separation of sister chromatids before the entire protein machinery is assembled [66].
3. **Metaphase** is characterized by correctly positioned sister chromatids, which are aligned at the equator of the bipolar spindle apparatus (Fig. 7.5). The anaphase-promoting complex (APC, see Sect. 7.3.4) facilitates the separation of the sister chromatids so that they can be pulled to the opposing poles of the future daughter cells.
4. **Anaphase** involves the split of sister chromatids away from each other. They are forced by the spindle apparatus towards the microtubule-organizing centre at the cell poles.

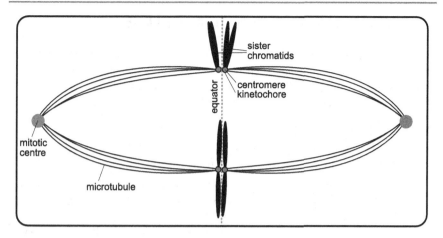

Fig. 7.5 Cell in metaphase of mitosis. Microtubules from mitotic centres attach with kinetochore around centromeres of sister chromatids that are aligned in the equator region of the cell. Bipolar spindle apparatus can then pull sister chromatids apart

5. **Telophase** exhibits loosening sister chromatids that start decondensation. Finally, a new nuclear envelope is assembled around each set of sister chromatids so that mitotic exit and cell division can follow.

Cell division separates the two nuclei by an encapsulating membrane. Synthesized proteins in the cytoplasm are assumed to be well mixed and thus equally distributed to both of the created daughter cells.

7.3.3   Find a Balance: Divide et cresce!

Division and growth of cells have to be well balanced to keep the right number of cells with the correct size. Although cause and effect in the relation between cell size and cell-cycle length are still debated, the following proportion is frequently observed:

$$\frac{\hat{t}_G}{\hat{t}_C} \propto \frac{n}{V},$$

with $\hat{t}_G$ being the average length of G_1-phase and $\hat{t}_C$ being the average duration time of a complete cell cycle and n being the number of cells and V being the overall volume of all cells.

However, the question remains whether the size of a cell is a mere consequence of growth rate and division frequency or independently regulated to meet the cell's physiological requirements [39].

7.3.4   A Molecular Cell-Cycle Set-up for Oscillations

We have seen the stunning capability of the cell-cycle machinery to up- and down-regulate individual components, such as cyclins (Fig. 7.2), repeatedly. The molecular requirements for a signalling network to produce oscillatory behaviour have been mentioned briefly (see Sect. 6.5.3) and apply for the regulation of the anaphase-promoting complex (APC) in various model systems [34].

APC is activated by binding Cdc20. This process is rather slow but facilitated through APC phosphorylation by M-phase complex of CDK1 with cyclin B (see Table 7.2). This complex re-establishes itself via Cdc25.

APC is a ubiquitin ligase that, once activated, induces securin destruction. Thereby, separase is unleashed that cleaves cohesin, which holds together sister chromatids that can subsequently be pulled apart.

Another target of active APC in complex with Cdc20 is cyclin B that is degraded and thus creates a negative feedback loop (see Sect. 6.5). Free CDK1 can then bind to APC in the late M-phase or the early G_1-phase of the next cell cycle. This complex is independent of APC phosphorylation and destructs cyclin A and cyclin B to prevent premature entry into the S- or M-phase. But CDK1 is phosphorylated by the complex between CDK2 and cyclin E in late G_1 to release APC and make it available for the next M-phase.

Timing plays an important role in cell-cycle regulation. Notably, the binding of Cdc20 to APC is rather slow, creating a sufficient delay of this molecular network motif (Fig. 7.6) required for oscillations [83].

7.3.5   The Cell-Cycle Programme

CDKs and cyclins mainly coordinate the cell cycle (Table 7.2). The process involves many checkpoints and regulatory components to control precisely timed progression

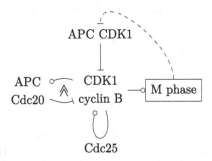

Fig. 7.6 The complex of CDK1 with cyclin B promotes M-phase progression and is established via Cdc25. Phosphorylation of APC by CDK1 promotes complex formation between APC and Cdc20. This complex degrades cyclin B thereby creating a negative-feedback loop. Complex formation between APC and Cdc20 is considerably slower than the degradation, creating a delay. APC forms also a complex with CDK1 in late M-phase that degrades cyclin B. Inhibition: ⊣. Catalysis: ─o

Table 7.2 CDKs and cyclins active in different cell-cycle phases in humans

	G_1-phase		S-phase	G_2-phase	M-phase
CDK	4,6	2	2,1		1
Cyclin	D	E	A		B

and prevent dysfunction. All these extracellular and intracellular conditions that are monitored influence the cell-cycle programme. Cells cycle only if sufficient growth factors or mitogens are available. They arrest in cell cycle, while DNA damage is repaired.

In the following, a heuristics of the cell-cycle programme is given as pseudo-code that the cell runs through during every division.

In G_1, the cell progresses if the external stimulus of growth factors or mitogens exceeds a certain threshold. CDK4 or CDK6 forms a complex with cyclin D that promotes gene expression via Rb phosphorylation and subsequent E2F release. This promotes also complex formation between CDK2 and cyclin E to establish G_1/S transition. Additionally, mTOR promotes growth by a metabolic shift towards protein synthesis.

In S-phase, the cell replicates its genome unless DNA damage is detected that needs to be repaired. Replication initiation is promoted by a complex of CDK2 or CDK1 with cyclin A.

In G_2, the cell waits to prevent premature entry into mitosis thus avoiding mitotic catastrophe.

In M-phase, the complex between CDK1 and cyclin B establishes binding between APC and Cdc20, which leads to the separation of sister chromatids. Finally, cyclin B is degraded to terminate the M-phase with cell division to start from anew.

```
%% G_1 phase
    if stimulus > threshold  % external growth factors or mitogens
        CDK4/6 + cyclin D = {CDK4/6 cyclin D}   % complex formation
        gene.expression = TRUE   % G_1/S transition
        CDK2 + cyclin E = {CDK2 cyclin E}    % complex formation
        mTOR.mediation = TRUE    % growth
    end
%% S phase
    while DNA.damage = TRUE       % cell-cycle arrest
        DNA.repair = TRUE
        DNA.replication = FALSE
    end
        CDK2 + cyclin A = {CDK2 cyclin A}    % complex formation
        DNA.replication = TRUE
%% G_2 phase
    wait
%% M phase
    CDK1 + cyclin B = {CDK1 cyclin B}    % complex formation
    APC + Cdc20 = {APC Cdc20}   % complex formation
    cyclin B = 0                % degradation
    return
```

Summary

- Different organisms serve as model systems to study the cell cycle because they are easy to grow and/or manipulate.
- There are several checkpoints at which the cell cycle halts until proper growth or maintenance of DNA integrity is ensured.
- The molecular programme for cell division is cycling as protein components of the machinery ("cyclins") are oscillating and other regulatory proteins ("cyclin-dependent kinases") depend on their activity on them.

Immunology

8

Contents

Introduction

The immune system is vital to human health and therefore a major field of biomedical research. *Immunis* in Latin originally means free from obligation or liability. The immune system can be understood as an intricate network or machinery that protects its host from foreign substances [24,57]. A key feature to exert this task is the discrimination between own and foreign entities.

The main subject of immunology is the immune system of our human body. Many phenomena in immunology carry human-related names such as recognition, memory and defence. Substances are recognized as self or non-self. There is a memory of the non-self substances and adaptation of the mounted immune response.

For all these phenomena to occur, a complex interplay is needed between the variable components that constitute our immune system. To cope with the diversity of immune responses, they are divided into the innate and the adaptive immune response. While the innate immune response relies crucially on macrophages, neutrophils and mast cells, it is linked by dendritic cells to the adaptive response that involves B lymphocytes and T lymphocytes. Adaptive immunity features a humoral response in addition to the cellular response.

Not only the mechanisms underlying these processes are in the focus of interest but also understanding of the immune system's malfunctions is important for curing disease. Errors in the immune system are referred to as autoimmunity, immunodeficiency or hypersensitivity and will mark the transition to cancer (see Chap. 9) towards the end of this chapter.

© Springer-Verlag GmbH Germany, part of Springer Nature 2022
L. Adlung, *Cell and Molecular Biology for Non-Biologists*,
https://doi.org/10.1007/978-3-662-65357-9_8

8.1   Components of the Immune System

The immune system consists of a multitude of components, which are organized in tissues as cells and molecules. The diversity of the immune systems' components reflects the plethora of causes of diseases, which have to be prevented. We as the host have to be protected from pathogens, which by definition have the ability to make us sick. Examples of pathogens are bacteria, fungi, parasites or viruses. The first line of defence is our skin and mucosal surfaces that line organs with contact with the outside world like lung and gut. Parts of our immune system recognize the intruding enemy specifically to fight it. The soldiers of this army of our immune system are the immune cells.

8.1.1   Sites and Non-Cellular Substances of Immune Reactions

A major home base of our immune cells is the bone marrow. Many immune cells originate from there. Hematopoietic stem cells (HSCs) in the bone marrow have the capacity to give rise to all immune and blood cells. White blood cells (leukocytes) form the majority of the cellular immune response with a myeloid and a lymphoid branch (see Sect. 8.1.2). Together with the thymus, the bone marrow is referred to as *primary* lymphoid organs because lymphocytes are derived there. The lymphatic system comprises vessels transporting fluids and cells through our tissues. This draining substance with all its components is referred to as **lymph**. Specializing lymphoid cells exert their tasks also in the periphery of our body. Lymph nodes, mucous, skin and spleen are therefore referred to as the *secondary* lymphoid organs in the periphery. Skin and mucous are the first physical barriers against pathogens. The mucous is a fluid covering our respiratory and gastrointestinal tract with neutralizing substances (e.g. IgA), acidic milieu (i.e. low pH) and enzymes (e.g. lysozyme) or even small peptides (e.g. defensin) that kill pathogens. Besides the cellular response, these chemicals, enzymes and peptides, there are other molecules, mainly antibodies (see Sect. 8.3.2) to fight pathogens. Antibodies are secreted to extracellular fluids, which constitute humoral immunity (Fig. 8.1).

All components of the immune system need to be well orchestrated to guard our body. Regulatory mechanisms and maturation dynamics were derived from the study of hematopoiesis, a process that gives rise to all our immune and blood cells.

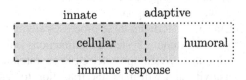

Fig. 8.1 There is an intersection in cellular components between innate and adaptive immune responses. In addition to cells, the adaptive immune response is mediated by humoral components, i.e. secreted antibodies

8.1.2 The Hematopoiesis Tree and Its Branches

Immune cells are a subset of the hematopoietic system that is rooted in hematopoietic stem cells. Because of their potency to give rise to several diverse cell types, they are called pluripotent. Cells arising from HSCs subsequently lose this potential for functional diversity. Progenitor cells become more and more committed to a certain fate. Upon maturation, cells have acquired a very precise function and cannot produce any other cell type. Every stage in this process is considered a compartment. There are compartments of stem cells followed by compartments of committed progenitor cells and finally compartments of mature cells.

The process of cellular reproduction within a compartment, involving iterated rounds of cell cycle (see Chap. 7), is called proliferation. Differentiation refers to cells acquiring specialized functions while changing their morphology or intracellular structures thereby entering a new compartment. The balance between proliferation and differentiation is important for homeostasis of the hematopoietic system. HSCs are on top of a hierarchy that is referred to as the hematopoiesis tree with "stem" cells and different branches of progenitor cells (Fig. 8.2). The arrows in the hematopoiesis tree indicate that differentiation is unidirectional and cells cannot differentiate back into a previous compartment. A more quantitative understanding could help to weigh and lengthen these arrows according to the frequency and duration of the fluxes. Many cells rapidly differentiating along a line would result in a thick but short arrow, whereas few cells slowly leaving a compartment would be indicated with a thin, long arrow.

In general, there is a differentiation influx parameter d_i into a compartment and a differentiation efflux out of the compartment with the rate constant d_o. An average cell

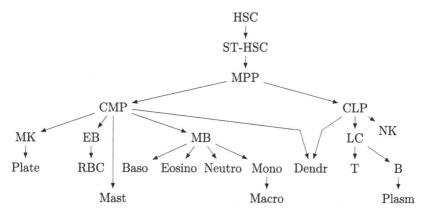

Fig. 8.2 Schematic overview of the hematopoiesis tree. Arrows are arbitrarily scaled. HSC: hematopoietic stem cell, ST-HSC: short-term hematopoietic stem cell, MPP: multi-potent progenitor, CMP: common myeloid progenitor, CLP: common lymphoid progenitor, MK: megakaryocyte, EB: erythroblast, MB: myeloblast, NK: natural killer cell, LC: large lymphocyte, Plate: platelet, Trombo: trombocyte, RBC: red blood cell, Baso: basophil, Eosino: eosinophil, Neutro: neutrophil, Mono: monocyte, Dendr: dendritic cell, T: t(hymus) lymphocyte, B: b(one marrow) lymphocyte, Mast: mast cell, Macro: macrophage, Plasm: plasma cell

Fig. 8.3 Cell x differentiates
into the shaded compartment
with parameter d_i and
progresses with d_o. The net
proliferation rate is
$(\lambda - \alpha) \cdot x$

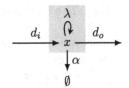

within the compartment proliferates with λ and has a probability α to die (Fig. 8.3).
Please note that for HSCs there is no influx, and for mature cells, there is no efflux.
These rates for proliferation, differentiation and death characterize the hematopoietic
system and its flux control. Genetic labelling of HSCs and subsequent progeny in a
mouse model allowed researchers to establish a quantitative framework to determine
those rates [18].

It was found that an adult mouse contains $\approx$5000 HSCs contribute to normal
hematopoiesis. And both, the HSCs as well as their subsequent compartment, the
short-term (ST)-HSCs operate near self-renewal. This means that the outgoing differentiation and death rate do not exceed proliferation and differentiation influx of
the ST-HSC compartment (Fig. 8.3):

$$d_i + (\lambda - \alpha) \leq d_o. \tag{8.1}$$

Additionally, the residence time of an average ST-HSC was estimated to be 330
days [18]. This is an exceedingly long time spent in this compartment given that
the expected lifetime of a mouse is around 2 years. Without stem or progenitor cells
being exhausted, those rates allow the hematopoietic system to rapidly replenish
compartments after depletion. Amplification is achieved by massive proliferation of
multi-potent progenitor cells that are the branching point for the myeloid and the
lymphoid lineage.

The myeloid lineage contains important cells of the innate immune response
(see Sect. 8.2), namely basophils, eosinophils, neutrophils, mast cells (collectively
called granulocytes), monocytes and macrophages and dendritic cells. Macrophages
mature from monocytes, and together with granulocytes and dendritic cells, they
belong to the group of phagocytes. Phagocytes are cells that eat up their enemies.
Besides other functions, e.g. during inflammation (see Sect. 8.2.1), macrophages act
as scavenger cells indiscriminately clearing cell debris. Granulocytes were named
after the multitude of diverse granules in their cytoplasm. These vesicles contain
enzymes and toxins to destroy microorganisms and parasites. Unlike macrophages,
granulocytes are not always present but are called into action if needed. Mast cells
are less well defined. They are known to be mainly involved in allergic reactions.
Dendritic cells take up pathogens, but in contrast to macrophages and granulocytes,
their main task is not clearance but presentation of antigens. These pathogen-derived
fragments received their name from the fact that *anti*bodies (see Sect. 8.3.2) are
generated by B lymphocytes in the adaptive immune response to antigen exposure.
Dendritic cells belong to the class of antigen-presenting cells (APCs). They exhibit
pathogen particles on their outer surface to expose them to naive T lymphocytes
that thereby become activated. Dendritic cells have arm-like structures to embrace

as many T lymphocytes as possible. Dendritic cells were named after those arm-like processes because they look akin to dendrites of brain cells.

During wound healing, platelets serve the purpose of clotting blood. Because they make blood thick and sticky, they are also called thrombocytes. Although red blood cells are not directly involved in immune reactions, they ubiquitously contribute to the overall number of cells in our body and thus the hematopoietic system. Recently, the number of red blood cells in an average male was estimated to be 2.5×10^{13}, which represents 84% of total cell counts [97]. Given the half-life of red blood cells being 120 days [46], our body produces more than 2 million mature red blood cells every second. These cells are needed because their vital task is to provide all our organs and tissues with oxygen.

Procedure

Erythropoiesis means the maturation of red blood cells from erythroid progenitor cells. Optimal control of proliferation and differentiation is important for erythroid progenitor cells to produce the right amounts of red blood cells in a given time. To define the stages of erythroid differentiation is a non-trivial task. Criteria for distinct subpopulations are morphology, expression of surface markers and functional assays. During maturation, erythroid progenitor cells undergo morphological changes as they shrink and eject their nucleus. The expression of the transferrin receptor CD71 and the glycophorin A-associated antigen GPA is regulated because these surface markers are implicated in iron metabolism. Iron ions are needed for haemoglobin synthesis, which is essential for oxygen-carrying red blood cells. Functional assays characterize haemoglobin content in a blood sample. Stages of erythroid progenitor cells are named burst- or colony-forming unit-erythroid cells, respectively, depending on how they grow in a semi-solid medium. To analyse thousands of single cells with respect to their morphology and surface markers, a technique called flow cytometry can be applied (Fig. 8.4).
◀

The lymphoid branch contains natural killer cells that belong to the innate immune response. The common lymphoid progenitor cell gives rise to a minor subpopulation of dendritic cells. As dendritic cells take up pathogenic particles and represent them to naive T lymphocytes, they form an important link between innate and adaptive immune responses. B lymphocytes and T lymphocytes exhibit an intricate mechanism of regulation that will be further explained for adaptive immunity (see Sect. 8.3).

Whether our hematopoietic system is correctly represented by this tree-like structure (Fig. 8.2) is currently under debate. Especially the discrete branching between the lymphoid and especially the myeloid lineage is challenged by most recent technological advancements. The progeny of individual barcoded cells can be traced *in vivo* after transplantation into mice [87]. Single-cell RNA sequencing suggested a robust maturation continuum based on gene expression profiles [86] rather than discrete gating based on the presence of surface markers in two-dimensional scatter plots (Fig. 8.4).

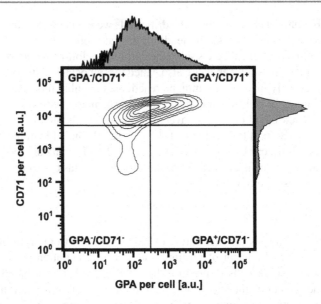

Fig. 8.4 Scatter plot of erythroid progenitor cells upon flow cytometry measurement. Single cells in a thin liquid stream pass laser beams. Recorded scatter light indicates the size and granularity of the individual cells. Surface markers can be stained so that fluorescent signals correlate with their abundance per cell. Gates are set to indicate cells positive for GPA or CD71, respectively. Every circle encloses 10% of a population of 38,475 cells

8.2   Innate Immunity as First Line of Host Defence

If a pathogen tries to invade our body, it first encounters multiple layers of the innate immune system. Mucous covers all wet surfaces of our body that are not protected by skin, such as eyes, mouth, lungs and stomach. Preventing these tissue layers (epithelia) from drying out stops cracking and thus keeps infectious agents away. The lubricious substance acts as a conveyor belt because it houses trillions of our bacterial tenants but flushes out pathogens. The importance of mucous is underlined by the fact, that dysfunctional mucous production is linked to disease: Mucoviscidosis aka cystic fibrosis. To further support tissue surface cleansing, white blood cells (leukocytes) are sent to spots of infection. Leukocytes are also recruited after wounding your skin because this first shield of your body is then permeable to pathogens.

8.2.1   Inflammation and the Complement System

If you wound your skin, the site of damage will start to hurt, swell and feel warm. This process is called inflammation. Microbes that breach the skin are recognized by macrophages, which engulf the bacteria and destroy them. This process is called phagocytosis. Besides eating up the enemy, macrophages release **cytokines**, which are messenger molecules to coordinate the immune response among the cellular components. A special class of cytokines, chemokines, attracts neutrophils and monocytes to exit the bloodstream and enter the local site of infection. Cells penetrating the blood vessels cause the characteristic pain of an inflammatory response. In addition, increased production of lymph flushes the invaders out of the tissue into nearby lymph nodes to trigger more specialized adaptive immunity. The elevated production of fluids and recruitment of immune cells cause swelling. Dilation of blood vessels and increased local blood flow to transport all the cellular components to the site of damage makes us feel heat, too.

Inflammation is enhanced by the complement system, which comprises a family of more than 20 plasma proteins. The complement system triggers a cascade of reactions to coat microbes with molecular patterns that are recognized by complement receptors on macrophages for phagocytosis. Proteins of the complement system not only attract macrophages but also cluster microbes together in a process called agglutination and lyse bacteria by punching holes through their cell wall and membrane. When the inflammation sustains, macrophages are supported by neutrophils, which are also capable of phagocytosis. The supply of macrophages is ensured by recruited monocytes that differentiate into mature macrophages at the site of local inflammation.

8.2.2 Pattern Recognition: Know Your Enemy

Precondition for any immune response is the recognition of pathogens. You have to know your enemy in order to combat them. Despite their diversity, many pathogens exhibit common motives on their surface that are essential and evolutionarily conserved. These structures are known as pathogen-associated molecular patterns (PAMPs). The presence of PAMPs is recognized by receptors that are called pattern-recognition receptors (PRRs). The number of different PRRs to detect pathogens is rather limited indicating the few core elements that discriminate pathogens from host cells. Detected components range from sugars and lipids (see Sect. 1.1) to DNA and RNA, in bacteria, fungi, parasites, viruses and its own host (reviewed in [5]). The major class of PRRs is Toll-like receptors (TLRs), named after the founding member Toll, which was initially identified in the fruit fly.

TLR4, for instance, a prominent member of the TLR family, recognizes lipopolysaccharides on the outer membrane of bacteria. Cell wall components of fungi can be detected by other TLRs, likewise plasma membrane anchors of parasites . Molecules of our own body should not but can trigger inflammation through TLRs, too. Heat shock proteins are secreted during fever upon infection and act in a pro-inflammatory way [105]. There are other molecules released by our own body cells, which are recognized by PRRs. They are called damage-associated molecular patterns (DAMPs), and as the name implies, they are released by stressed cells, which are dying without an infection. Hereditary material of bacteria or viruses can also be detected by TLRs. Unmethylated cytosine-guanine dinucleotides in a particular base context are characteristic of genomic DNA of bacteria and can therefore be recognized by TLRs. The RNA, no matter single- or double-stranded, of viruses can also elicit an innate immune response via TLRs because there is no "naked" single-stranded RNA (mRNA is, for instance, capped, see Sect. 3.2.2) nor double-stranded RNA under normal conditions in human cells. To study an anti-viral response, a synthetic analogue of double-stranded viral RNA, polyinosine-deoxycytidylic acid (poly I:C), is often used [8]. Activated TLRs trigger an inflammatory response by the expression of inflammatory cytokines, such as interferon (IFN)-α. TLRs detecting DNA or RNA are not located at the plasma membrane but at intracellular vesicles of phagocytotic cells because the enemy has to be processed first before its inner material can be exposed. Dendritic cells that process pathogens link the innate to the adaptive immune response. They migrate from the local site of infection to lymph nodes to activate naive T lymphocytes.

8.3   Adaptive Immunity as Ability to Strike Back

Processed pathogens have to be presented to the cellular components of the adaptive immune system to retaliate. Dendritic cells pick up antigens at the local site of infection and migrate to regional lymph nodes to activate naive T lymphocytes. Differentiation and proliferation give rise to functional forms of T lymphocytes, among them T helper cells, which help B lymphocytes to mature and produce antibodies in response to antigen exposure. The "T" in T lymphocytes stands for the thymus where they mature even though they originate from the bone marrow. B lymphocytes originate from and mature in the bone marrow, but the "B" stands for bursa of *Fabricius*, a lymphoid organ in birds from which they were first derived. Both, T lymphocytes and B lymphocytes, carry antigen receptors.

Antigens exhibit immense diversity, which is encountered by a possible number of 10^{15} different antigen receptors. It is impossible to maintain the presence of so many different receptors all the time, especially because it exceeds the total number of lymphocytes by a factor of ten thousand. Instead, only the cell that carries a receptor that recognizes the exposed antigen should be amplified in number. It has to be ensured that the receptor does specifically recognize the pathogen-derived antigen but does not cross-react to cells or fragments of our own body. High numbers and high specificity of so-called effector cells in the adaptive immune response are achieved by the process of clonal selection.

8.3.1   Clonal Selection of Special Forces

Lymphoid progenitor cells (Fig. 8.2) give rise to a large number of lymphocytes, each with an individual antigen receptor. All progeny of one of those cells will carry the same receptor, which is therefore a **clonal** population. If the receptor reacts to particles of the own body (i.e. a self antigen), it is discarded. Induced cell death of self-reacting lymphocytes is referred to as clonal deletion. Only if the receptor recognizes

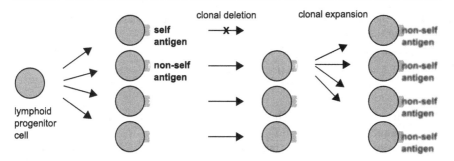

Fig. 8.5 Schematic overview of clonal selection for specific antigen-receptor cells. A lymphoid progenitor cell gives rise to different cells, each carrying an individual type of antigen receptor. Cells recognizing self antigens are deleted to prevent an auto-immune reaction. Cells specifically recognizing the non-self antigen of the foreign pathogen expand in number to eradicate the non-self antigen

specifically the antigen derived from the pathogen (i.e. a non-self antigen), the clone proliferates in a process called clonal expansion. The progeny carrying all the same receptor, which reacts specifically to the foreign antigen (Fig. 8.5). Thus, specialized T lymphocytes form functionally distinct sub-populations and B lymphocytes give rise to plasma cells that can produce the antigen receptor and secret it as an antibody.

8.3.2   Antibodies

Antibodies are an important weapon of the immune system. They detect pathogen-derived antigens and neutralize them or activate further cells of the adaptive immune response. These two functions of recognizing an antigen specifically and affecting different cells in turn are represented in the Y-shaped structure of an antibody. Each of the two upper arms of the Y binds to an antigen, while root at the bottom interacts with cells. Antigens are very variable and so are the antibody regions detecting them. The same two regions of an antibody that detect two antigens of the same type are called **variable regions** because they differ from antibody to antibody, each being specific for a single type of antigen. On the other end, there is a limited set of cell types of the adaptive immune system interacting with the antibody. The antibody's effector function is thus exerted by a **constant region**.

The Y shape implies a vertical symmetry axis for antibodies. They are composed of two identical heavy chains linked by disulfide bonds and two identical light chains on the outer arms linked to each heavy chain by disulfide bonds, too. Heavy and light refer to the approximate molecular weight of 50 kDa and 25 kDa, respectively. The immense diversity between the variable regions of the individual antibodies is created by shuffling of gene segments by means of DNA recombination (see Sect. 2.4.2) in the different cell clones. The DNA stretches that are shuffled are the **V**ariable gene segment, the **D**iversity gene segment (only in the heavy chain) and the **J**oining gene segment. The process is therefore referred to as **VDJ recombination**. In contrast to the vast number of variable regions, there are only five different types of constant

Table 8.1 The five different kinds of heavy chains determine the class of the antibody, called immunoglobuline (Ig)

Heavy-chain type	μ	γ	α	ϵ	δ
Immunoglobulin class	IgM	IgG	IgA	IgE	IgD

regions for the heavy chains. They correspond to the different classes of antibodies called immunoglobins (Table 8.1)

Antibodies are secreted by matured B lymphocytes in huge quantities, which makes them the major component of the humoral (i.e. non-cellular) immune response and suitable for being studied biochemically. However, antibodies are not the only molecule capable of antigen binding. Both, the B cell receptor (BCR) and the T cell receptor (TCR), were selected to specifically bind antigens. These receptors are structurally and functionally related but not identical (Fig. 8.6). While BCRs exhibit a Y shape, TCRs are rather I-shaped. Every BCR carries two antigen-binding sites whereas each TCR can only bind a single antigen. The specific BCR can be bound to the membrane of a B lymphocyte, and, the very same receptor, can be secreted as an antibody. TCRs are always integrated into the membrane of T lymphocytes and can only recognize antigens that are presented by antigen-presenting cells (APCs).

8.3.3   Regulation of T Cells for Killing, Help, Memory and Balance

T lymphocytes play a very central role in adaptive immunity. Naive T lymphocytes can mature into cytotoxic T cells, which acquire the ability as effector cells to kill the pathogen. T helper cells constitute another sub-population of T lymphocytes, which helps B lymphocytes to mature into plasma cells for massive antibody production. Besides, so-called regulatory T cells prevent the immune system from overshooting. How are these diverse functions orchestrated? Cytokines are important soluble factors secreted by cells to coordinate the innate immune response (see Sect. 8.2). Adaptive immunity relies on direct interaction between cells via receptors on the surface of immune cells. These receptors are called clusters of differentiation (CD).

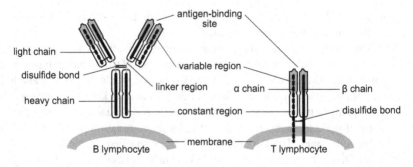

Fig. 8.6 Structure of an antibody as secreted form of a B cell receptor and the membrane-bound T cell receptor

They are traditionally used to classify sub-populations of the hematopoiesis tree. More than 400 different CD molecules have been described in humans [33]. The decision of whether a cell belongs to our body or not is nonetheless binary. To discriminate between self and non-self, there is another important type of surface marker: the major histocompatibility complex (MHC).

Type-I MHC molecules are expressed by all nucleated cells of our body. Except red blood cells, all our blood cells identify themselves by carrying the MHC-I molecule on their surface. The human MHC molecules are therefore referred to as human leukocyte antigen (HLA). The name for **MHC** molecules was derived from the fact that they define **tolerance** to tissues upon transplantation. If tissue is compatible, the subtype of the major histocompatibility complex matches between donor and recipient. As there are minor MHC antigens that could evoke an immune response, suppression of the immune system is needed to avoid rejection of the tissue, even if the MHC-I subtype matches. Cells lacking MHC I on their surface are eliminated by natural killer cells as part of innate immunity. MHC molecules are no real receptors but rather membrane-embedded carrier molecules, which expose certain antigens to surrounding cells. Antigens displayed by MHC I are derived from cytoplasmic particles of the cell upon viral infection. Antigens presented by MHC II are derived from extracellular particles that were taken up and processed by antigen-presenting cells (APCs) in intracellular vesicles.

While infected cells are carrying the viral particle loaded to MHC I have to be eliminated to prevent the spread of infection, APCs need to be further supported and orchestrated in fighting pathogens. These tasks are exerted by two distinct subsets of T lymphocytes. The killing is done by **cytotoxic T cells**, and the mediation of the adaptive immune response is performed by **T helper cells**. To prevent confusion between the MHC molecules and the subsets of T lymphocytes, co-receptors are expressed on their surface that corroborate correct detection. Besides the T cell receptor on the surface of cytotoxic T cells detecting the viral antigen presented by MHC I, a **CD8** molecule binds MHC I to confirm the killing. In short, those cytotoxic T lymphocytes expressing the CD8 molecules on their surface are referred to as cytotoxic CD8 T cells. And apart from the T cell receptor detecting antigens presented by the MHC II, there is a **CD4** molecule on the surface of T helper cells binding MHC II to confirm the release of cytokines, e.g. for activation of B lymphocytes and antibody production.

Some of the matured T and B lymphocytes remain after the pathogen has been defeated thereby forming the basis of immunological memory. These **memory cells** can re-activate the adaptive immune response much quicker in case of another encounter with the then "known" pathogen. The adaptive immune system thus exhibits a memory because it remembers previous attackers. In theory, the property of a system's output depending on the history of input is known as hysteresis. In the context of adaptive immunity, this means that the strength of the immune response rises with the increasing strength of first pathogen exposure. Once exposed, the adaptive immune system remembers the pathogen and mounts a strong response even if the pathogen exposure declines the second time (Fig. 8.7). Hysteresis was in fact

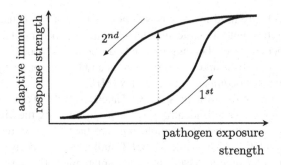

Fig. 8.7 Adaptive immune response strength versus pathogen exposure strength. For the 1st exposure, T lymphocyte activation rises with increasing antigen stimulation. For the 2nd exposure, immune response remains strong, even if pathogen levels sink. The curve forms a hysteresis loop that depicts memory. For the same pathogen exposure, the 2nd response adapted as compared to the 1st (dotted arrow)

shown with concentration of activated T lymphocytes versus antigenic stimulation level [17].

If receptors are detecting self antigens, death of the cell carrying this receptor can be induced (see Sect. 8.3.1). Apart from this self-induced cell death, there is a population of functional T lymphocytes that suppresses immune responses. These regulatory T cells ("Tregs") exist as mature forms in the periphery of our body and exert dominant-negative control over self- or hyper-reactive T lymphocytes to balance the immune response (i.e. homeostasis). If regulatory T cells are functionally deficient, chronic auto-immune diseases occur [92].

8.3.4   (Auto-)immune Disease

Our immune system involves the following features:

- diversity of the repertoire of antigen receptors,
- specificity for pathogen-derived antigens,
- discrimination between self and non-self antigens and
- memory of past antigen exposures.

Mechanisms have to be well balanced for a reaction with correct timing, specificity and strength. Too weak immune response will let the pathogen spread and an over-reaction exhausts resources, both harmful to our body. To sum up, diversity of the antigen receptors is generated by genetic shuffling called VDJ recombination, while specificity is achieved by the clonal selection of antigen-receptor cells (Fig. 8.5). After the clonal selection, discrimination between self and non-self is additionally ensured by MHC molecules identifying our body cells or displaying antigens of invaders. Some T and B lymphocytes that orchestrate cellular and humoral immune response persist after the initial exposure to strike back more efficiently at the second attack. This is a strong call for vaccinations to prevent infections. Because if the body is exposed to dead and therefore harmless particles derived from a pathogen, memory cells specifically recognizing these antigens can form and provide a long-lasting protection against the disease.

Problems occur if the process loses control. In cases of autoimmune diseases, components of ourself are falsely detected and fought as the enemy. Our own cells then make us sick, and the host itself becomes the pathogen. An example of an autoimmune disease is type I diabetes. In this disease, over-reactive CD8 T lymphocytes and B lymphocytes lead to the destruction of insulin-producing cells in our body. Over-reacting lymphocytes can also give rise to lymphoma, a cancer of the lymph nodes. On the other hand, natural killer cells and cytotoxic T cells could be used to specifically eradicate cancer cells. But immediate questions are: What is cancer, and how to detect, characterize and combat it? These questions will be addressed in the final chapter.

Summary

- Cellular and non-cellular (i.e. "humoral") components constitute the immune system to defend an organism from disease-causing entities.
- All immune cells are highly specialized in their function and arise from common and shared progenitors all originating back to hematopoietic stem cells.
- Whereas innate immunity is about pattern recognition of common pathogens, adaptive immunity is triggered against novel invasions through selection of highly reactive non-/cellular components.

Cancer

9

Contents

Introduction

Cancer is one of the most devastating diseases of our time. It is a global challenge to find new ways and improve existing approaches to prevent, diagnose and combat cancer. Oncology is the name of the subject that is dedicated to understanding the cause and the progression (i.e. the "aetiology") of the disease(s). The biggest hurdle in fighting cancer is that there is no single enemy. Cancer represents a multitude of diseases with a huge variety of modes of action. Therefore, we have to join forces from different fields of medicine, physics, biology and mathematics: surgery, pharmacology, radiology, molecular biology and dynamic modelling.

Cancer represents the final chapter of this book because we need to consider all aspects of modern biology to understand the basics of cancer research, from the change of cellular architecture and the deregulation of gene expression to signal transduction, cell cycle control and the role of the immune system. Despite the complexity of cancer, there is one characteristic that all different types of cancers share to a certain extent: The deregulation of cell proliferation. Proliferation is phenomenologically a straightforward process: One cell becomes two in a given time. This duplication involves DNA synthesis, cell growth and division. In normal, healthy cells under physiological conditions, proliferations are tightly controlled. If this regulation is lost, for various reasons, healthy cells transform into cancer cells that proliferate in an uncontrolled manner (Fig. 9.1). Erroneous DNA synthesis, excessive cell growth and frequent cell division lead to the formation of tumours and metastases.

© Springer-Verlag GmbH Germany, part of Springer Nature 2022   103
L. Adlung, *Cell and Molecular Biology for Non-Biologists*,
https://doi.org/10.1007/978-3-662-65357-9_9

Fig. 9.1 Cancer is characterized by uncontrolled cell proliferation. While a healthy cell doubles only once in a given time, a cancer cell grows and divides much faster in the same time to form tumours and metastases

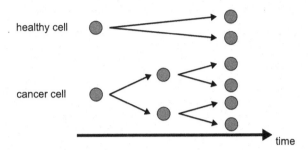

There are many physiological anti-cancer mechanisms that keep cell proliferation under control. Once a former normal cell escapes and is transforming into a cancer cell, drug interventions can help to constrain the tumour in its size or localization inside the body. But cancer cells divide and therefore evolve rapidly. They develop resistance against therapeutic approaches. Besides, interfering with a generic process such as proliferation can also have severe side effects and harm healthy body cells. But recent discoveries regain us hope that integrated and personalized approaches may lead to a life without cancer in the future if we combine our expertise from the diverse areas of modern biology.

9.1   Different Types of Cancer

Cancer is diverse. There is not only a single type of cancer. There is lung cancer and breast cancer, colon cancer and blood cancer, tumours in the brain and the prostate, just to name a few. But even within one and the same organ, there can be different subtypes of cancer, at subsequent stages of severity of the disease. However, cancer cells arise from normal cells, tumours form within healthy tissues. Given this origin, there must be some shared characteristics. Cancer was named after the brown crab *Cancer pagurus* because the shape of tumours in Ancient Greece reminded people of this animal [76]. According to this comparison, the body of crab refers to the **primary tumour**, which is its core, its site of origin. The distant legs and claws represent ensembles of cancer cells spread to other places, called **metastases**. Tumours that do not grow out, form metastases and invade other sites, but stay local, are called **benign**, whereas tumours that invade other tissues are called **malignant**. If tumours originate from epithelial cells, sheets that line inner and outer surfaces of tissues, they are called **carcinomas**, which make up 80 % of all human cancers. Within the carcinomas, one can discriminate between squamous carcinoma, originating from protective epithelial cell layers, and adenocarcinoma, generated by epithelial cells that secrete substances to inner cavities. As such there can be both, squamous and adenocarcinoma, in the lung, depending on the originating cells.

A major group of cancers, which do not derive from epithelial cells, are the **sarcomas** that stem from connective tissue, consisting largely of so-called "mesenchymal cells". Sarcomas constitute only 1 % of tumours. An example of sarcomas is liposarcoma that affects fat tissue. Tumours of the hematopoietic system are called

leukaemia if they refer to myeloid cells or **lymphoma** in case the lymphoid lineage is concerned. The last major group of cancers relates to neuroectoderm cells and thus the brain. This group of cancers involves gliomas, glioblastomas, neuroblastomas, schwannomas and medulloblastomas. Again, the name depends on the cell type of origin or where are they localized.

There is a challenge to this dogma, because (cancer) cell types and their localizations can change during the generation of a tumour. This feature is called "**transdifferentiation**" and means change in cell type. Most prominently, epithelial cells can become mesenchymal cells during a process, which is referred to as epithelial-mesenchymal transition (EMT). In simple terms, a surface cell is becoming a cell of connective tissue properties, which is important for the cancer cell to invade neighbouring tissues [11].

9.1.1   Phenotypic Tumour Development

Cancer is diverse. An example of some of the vocabulary introduced above can be the following common path of colon cancer development. The type of cancer is routinely defined based on the site of its first occurrence, i. e. the primary tumour in case of colon cancer in the colon, which is the last part of the gut before the rectum. Other tissues can be affected during the course of the disease, e.g. metastases in the liver. It starts with **hyperplasia** of the epithelial cell layer in the colon. This means that all cells increase in numbers while still looking healthy. Once cells stop looking healthy, the phase of **dysplastic** growth is entered. How does an unhealthy cell look like? An example is an increased ratio between the volume of the nucleus and the volume of the cytoplasm. Therefore, a biopsy is taken, and from a thin slice of the epithelial layer of the colon, DNA is stained for histology ("the study of tissue").

This abnormal cell growth will soon be macroscopically visible by the naked eye in the form of a polyp. Since the epithelial cell layer in the colon is capable of secreting substances, the polyps can be referred to as adenomas. Yet they are no adenocarcinomas, because the polyps do not penetrate through the basal membrane on the tissue and thus do not invade underlying tissue. Adenomatous polyps are benign.

Once the abnormal cells break through the basal membrane and invade other tissues, they are referred to as (malignant) **carcinoma** cells. The process of abnormal cell growth in disparate places such as epithelial cells in the underlying stroma of the colon is called **neoplasia**, because new tissue types are being formed (Fig. 9.2). Some people use the term "neoplasm", which is the product of neoplasia, collectively for malignant tumours. A malignant tumour in the colon is certainly colon cancer, potent of undergoing EMT and forming metastases in other organs.

In general, a **neoplasm** can also be benign. There are even cases, in which almost normally looking cells are found in disparate places, a phenomenon called "metaplasia". This example highlights the **plasticity** in the context of cancer development. There is not a single way of tumour formation. For the sake of the example above, the most common way of cancer generation can be depicted as follows [109]:

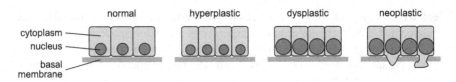

Fig. 9.2 Different growth stages of a cells from benign (left) to malignant (right)

$$\text{normal} \rightarrow \text{hyperplastic} \rightarrow \text{dysplastic} \rightarrow \text{neoplastic} \rightarrow \text{metastatic}$$

Finding the origin of tumour cells is an important aspect to eradicate cancer. In theory, the majority of the tumour mass could relate back to a single cell or multiple different cells. Individual cells and their progeny are called "clones", because they are genetically very similar and link back to a common ancestor cell. So when tumours are genetically homogeneous, it can be assumed that most of the tumour cells originate from a single clone, and the tumour itself is thus monoclonal. If there is more than a single dominating subset of tumour cells, the tumour is heterogeneous and polyclonal (Fig. 9.3).

Tracing experiments have shown that most of the tumours are indeed of monoclonal origin. So a single cell outcompetes the others and soon dominates tumour mass. But since cancer cells in general are rapidly dividing and evolving, new subsets of tumour cells can emerge that deviate genetically and phenotypically from the original clone [13], again highlighting cancer plasticity, which represents a major challenge for the development of effective medications. How come, a disease such as cancer is so diverse?

9.1.2   Causes of Cancer Development

Cancer is a multifactorial disease. The factors that can—directly or indirectly—lead to cancer include:

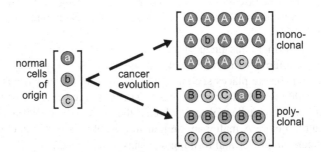

Fig. 9.3 Normal cell clones start to multiply unrestrictedly, cancer starts evolving. While a monoclonal tumour is dominated by a single-cell clone, polyclonal tumours exhibit more than one dominating clone

- **Genetics** The presence or absence of certain gene variants can alter the risk to develop certain types of cancers. An example is the gene BRCA1, which encodes for the breast cancer type 1 susceptibility protein. Mutations in BRCA1 represent a heritable genetic predisposition to develop breast cancer, which can be tested for [70].
- **Life-style** Whether we expose ourselves to cancer-inducing ("carcinogenic") substances, such as tobacco smoke, is a question of lifestyle. On the other hand, physical activity and a well-balanced diet can serve as preventive measures lowering the risk to develop cancer [55].
- **Environment** The effect of UV light from the sun, which increases skin cancer susceptibility is well documented [30]. Besides, other environmental factors such as pollutants can contribute as one out of many factors to the development of cancer.

How much do these individual factors weigh in the formation of malignant tumours is hard to quantify. There are interdependencies between genetics, lifestyle and environment. You may live in an area with high UV exposure but your lifestyle features extensive use of sun protection or interior activities. Or your genetics exhibit a mutation in the melanocortin 1 receptor (MC1R) gene, which renders you even more sensitive to UV light [30]. These examples highlight the complex interplay of multiple factors. Each factor alone can either increase or decrease the risk to develop cancer but they also affect one another. Adapting your lifestyle aids cancer **prevention**.

UV light can cause cancer because it induces mutations. The consequences of those changes in the genetic code depend on *where* they occur. If mutations affect the DNA repair machinery (see Sect. 2.4.2), the risk increases that errors are propagated. The effect of mutations in other genes depends on the role of those genetic regions in homeostasis under physiological conditions.

9.2   *Pro et Contra* Cancer

There is a fine balance between pro- and contra- (or anti-) carcinogenic signalling. Genes that are linked to the promotion of cell growth and multiplication are the origin of tumour formation, because if they are hyperactive, proliferation becomes unrestricted. Genes that have been linked to the generation of cancer are called **oncogenes**. A prime example for an oncogene is Ras. Ras stands for "Rat sarcoma" and was discovered as a leukaemia virus gene capable of transforming normal cells from rodents (e.g. rats and mice) into cancer cells [110]. The original ("wild-type") version of the gene encoding Ras is not yet considered an oncogene because it requires the alterations present in the viral gene to become hyperactive. Therefore, the original Ras is referred to as a *proto*-oncogene. Mutations in the original Ras gene ortholog were soon after found in human tumours. These mutations render the protein hyperactive and lead to uncontrolled proliferation (see Sect. 7.2.3). Negative regulators of oncogene signalling can suppress tumour formation. These genes are **tumour suppressors**. Prime examples are the dual-specificity phosphatase (DUSP)

and the phosphatase and tensin homolog (PTEN), which are both negative regulators of Ras signalling. Other tumour suppressor genes are involved in DNA repair
(see Sect. 2.4.2), such as BRCA1. Through various ways, mutations rendering tumour
suppressors dysfunctional contribute to erroneous cell-cycle control, continuous proliferation and tumour formation.

9.2.1   Immortality and Programmed Cell Death

Once cells proliferate continuously, they will steadily do so in a dish in the lab and
can be utilized as "immortal" cell lines (see Sect. 7.1.1). Immortal does not mean
that individual cells never die, it just means that cells never cease to proliferate,
and the population is maintained. In other words, these cell populations can become
super old. How is this achieved? A common feature of tumours and immortal cell
lines seems to be an overexpression of telomerases (see Sect. 2.4.1). Telomerases can
counteract the shortening at the ends of chromosomes during DNA replication. A lot
of telomerase activity is needed when cells replicate their DNA rapidly to progress
through the cell cycle repeatedly [15]. Hyperactive telomerases in cancer cells are
of course bad for our health. In normal cells, hyperactive telomerases are associated
with healthy ageing and longevity.

On the contrary, a way to prevent a cell from proliferating uncontrollably and
forming malignant tumours is the induction of programmed cell death. This physiological process is called **apoptosis** and has to be tightly balanced for the maintenance
of tissue integrity. Too little killing can cause malignant outgrowth, whereas too much
killing can affect healthy cells and cause lesions. Apoptosis is controlled in two ways:

- **Extrinsically** through ligands from the outside of the cell (to be killed). Extracellular death-inducing signals include tumour necrosis factor (TNF) and the
  TNF-related apoptosis-inducing ligand (TRAIL).
- **Intrinsically** through intracellular stress during DNA replication or during ATP
  synthesis at mitochondria.

Both, extrinsic and intrinsic signals serve as **upstream regulators** of apoptosis cul-
minating in the activation of **downstream effectors**. Most prominent among those
effectors are proteases of the caspase family. Proteins have to be degraded in the
cell to be killed, and caspases do that job. Apoptotic cells are full of granules and
vesicles as they are partially digesting themselves. Remaining chunks are eaten up
by neighbouring cells some of them specialized phagocytes from the immune system
(see Sect. 8.2.1).

There are also negative regulators of apoptosis that prevent programmed cell
death, such as the Bcl2 family. Most of them function through binding and suppression of pro-apoptotic triggers at the outer membrane of mitochondria. If integrity
of mitochondria is preserved, there is no release of cytochrome c, and in turn no
activation of caspases. Without the activation of downstream effector caspases, programmed cell death is prevented. Cranking up the expression of negative regulators

of apoptosis is therefore commonly observed in cancer cells—no wonder Bcl2 is named after B cell lymphoma 2.

There is a more splatter-like version of cell death called "**necrosis**". While the corpses of apoptotic cells are fragmented and eaten up by neighbouring cells, necrotic cells swell up and explode thereby scattering their inner contents to the local surroundings of the tissue microenvironment. Released substances can attract additional immune cells with various (also promoting) consequences for tumour growth.

If cells are not dying but just partially digesting themselves in intracellular lysosomes due to the lack of nutrients as energy sources, the process is called "**autophagy**"—self-eating. Conceptually, autophagy also involves upstream regulators and downstream effectors. These molecules are not the same as for apoptosis, but there is some notable crosstalk [80]. While a starved tumour (i.e. deprived from nutrients and growth factors) cannot hyperactively proliferate, it may become therapy resistant [112].

9.2.2 Molecular Complexity Underlying Cancer

The concept of proto-/oncogenes and tumour suppressors as well as the regulation of the different flavours of cell death highlight the molecular complexity underlying cancer formation and its counteraction. There is a gradual process of a normal cell transforming into a cancer cell ultimately giving rise to a malignant tumour. A single gene, such as mutated Ras or BRCA1, can be sufficient to radically change multiple properties of a cell and its progeny. Altered cell size, shape, metabolism and division times indicate loss of proliferation control leading to the accumulation of tumour mass.

Despite this observation, cancer does not fit the one-gene-one-disease paradigm. An example for a disease, which is caused by a mutation in both alleles of a single gene, is cystic fibrosis. The affected gene in cystic fibrosis is CFTR. Cancer is not a single disease, it is multiple diseases, and it is not necessarily caused by the malfunctioning of a single gene. Instead, multiple genes can be involved; in fact, entire networks of genes [63]. And it is not always the same gene(s) that initiate the process. Mutated BRCA1 can lead to tumour formation, but by itself, it is not necessarily the cause of breast cancer. By hampering DNA repair, BRCA1 mutant most likely contributes to erroneous other genes through which entire molecular machineries start to go wrong leading ultimately to uncontrolled cell proliferation.

Due to tumour type diversity and complex gene networks involved, researchers have long been striving to identify common features - the so-called hallmarks of cancer—for the development of therapeutic interventions tackling those hallmarks.

9.2.3 Hallmarks of Cancer

There are different hallmarks of cancer, which are commonly found during the establishment of a tumour [43]. As noted above, there are many individual exceptions to those rules. Regardless, the hallmarks of cancer can be seen as the common denominator of cancer:

⊞ **Pro proliferation signalling**
The information to grow and divide needs to be sustained for a tumour to form. This is achieved through positive feedback loops of mitogen signalling stimulating the source and neighbouring cells to progress in cell cycle (see Sect. 7). Pathways that induce proliferation are constitutively (i. e. "constantly") active through mutations in the MAPK pathway. In addition, negative regulators such as DUSP become dysfunctional through mutations. All these events lead to excessive pro-proliferative signalling.

⊟ **Contra growth suppression**
There are several safety measures that keep a cell from proliferating too fast. Overactivation of pro-proliferative signalling can trigger a cellular growth arrest known as "senescence" [25]. Cancer cells adapt to sub-maximal mitogen signalling for optimal proliferation avoiding the induction of growth arrest. Molecularly, this can be achieved by a loss of function of tumour suppressors such as Rb or p53.

⊞ **Pro invasion and metastases**
If cells grow too densely, further proliferation is prevented by a phenomenon called "contact inhibition". In simple terms: Cells stop to multiply once they start touching each other. Cancer often overcomes such inhibition, for instance by interfering with anchoring junctions through cadherin (see Sect. 1.4.2), thereby also loosening the extracellular matrix. In such cases, the tumour suppressor NF2 is mutated [84]. Another important factor in this context is the transforming growth factor beta (TGF-β), which is usually responsible to shut down cell proliferation. In the cancer context, TGF-β is known to induce epithelial-mesenchymal transition (EMT) as a precondition for metastases.

⊟ **Contra senescence and cell death**
For cancer cells to become immortal, two barriers to proliferation need to be overcome. Senescence and apoptosis. Upregulation of telomerases [15], and anti-apoptotic genes such as Bcl2 [44], have been named as two measures for tumour cells to proliferate continuously. It is worth noting that expression levels of telomerases can vary dynamically throughout the process of cancer development as pre-malignant lesions in the human breast exhibited reduced telomerase expression levels and shortened chromosome ends [22].

⊞ **Pro blood vessel formation**
For a sufficient supply of the growing tumour with oxygen and nutrients, blood vessels need to penetrate the malignant tissue. The process of blood vessel formation is known as "angiogenesis", which mostly happens during the development of the embryo and in the adult organism predominantly in wound healing, besides tumours. Some human cancers such as those in the kidney exhibit a very high

density of newly formed blood vessels [115]. Of note, already pre-malignant tumours promote the formation of blood vessels [74], which was first believed to be a property of macroscopically visible tumours only.

Not all the hallmarks are necessary or even essential to the generation of cancer. Tumour blood supply alters local metabolic properties within and around the malignant tissue, which changes also the immune landscape. Such overarching principles are commonly observed:

Tumour metabolism
For a tumour cell to proliferate continuously, the cellular metabolism needs to be reprogrammed. Normal cells use glucose as an energy source and first process it to pyruvate in a reaction chain called "**glycolysis**" (i.e. the degradation of glucose). If enough oxygen is present, normal cells then paste pyruvate to mitochondria (see Sect. 1.3) for oxidation and further energy production in the form of ATP. Conversely, cancer cells mostly stick to glycolysis even in the presence of sufficient amounts of oxygen. What seems like an inefficient energy production strategy turns out to be advantageous to fuel tumour growth. Cancer cells adapted to efficiently take up glucose from their surrounding through high-affinity glucose transporters such as GLUT1 [52]. Glucose uptake by the tumour is so high that it can be visualized by a positron emission tomography (PET) scan with a radioactively labelled form of glucose.

Once cancer cells start to hyperactively proliferate, oxygen supply in the tissue might become limiting, which is a problem for normal cells, because they rely on the oxygen for their energy metabolism but cancer cells are already adapted and more independent from oxygen. In addition, low oxygen conditions result in stress responses and apoptosis of normal cells but therapy resistance of metabolically adapted cancer cells [51].

Of note, tumour metabolism first and foremost shapes the (**tumour**) **microenvironment** of the respective tissue, which is referred to as the local surrounding spanning only a few layers of cells around the growing tumour. This spatial vicinity is directly altered by the cancer cells, e.g. in terms of glucose or oxygen concentration or abundance of signalling molecules.

Tumour immunology
The metabolic properties of a tumour render the malignant tissue a niche for immune cells. In fact, human tumours are infiltrated by cells of both the innate and the adaptive arm of the immune system to various degrees [36]. Ideally, for the human host, immune cells recognise cancer cells as something foreign and kill them, e.g. through cytotoxic CD8 T cells (see Sect. 8.3.3). Fortunately for the tumour, it has evolved from normal cells of the body and therefore escapes immune surveillance or weakens the immune response by means like exhaustion of adaptive T cells [111]. Among the innate immune cells, tumour-associated macrophages have been described, interacting with cancer cells, reciprocally stimulating one another thereby fuelling metastatic dissemination of breast cancer [113].

Intuitively, promoting inflammation of the tumour should aid the immune-mediated eradication of cancer cells. But the example of the macrophages above exemplifies that tumour cells recruit particularly cells of the innate immune system to exploit them, e.g. as sources of growth factors and nutrients like in prostate cancer [72].

These examples highlight the dual role of the immune system, which can both suppress and enhance cancer development.

Tumour plasticity

The presence of immune cells within the tumour underlines the fact that malignant tissues are no uniform array of cells. Even among cancer cells exists a considerable **heterogeneity**. Reasons for those differences between cancer cells within the same tumour are mostly due to genetics and the microenvironment [73].

- **Genetics as a reason for tumour plasticity**: In general, chromosomal instability of cancer cells is achieved through mutations in key mediators of genomic integrity such as BRCA1 involved in DNA repair, or mutations in the tumour suppressor p53, the original "guardian of the genome" [65]. Systemic defects in genome maintenance are key for cancer cell-to-cell variability and tumours to develop.
- **Microenvironment as a reason for tumour plasticity**: The physical and chemical properties at the core of a tumour are very different from the properties at the outer margins. The proximity around blood vessels is different from hubs of cytokine-secreting immune cells. This diversity gives rise to cancer cells of various shades, e.g. with respect to their metabolic capacity [96].

Consequence of cancer cell heterogeneity is plasticity, which means the ability to adapt. If there are many different cancer cells within a tumour, the chance is high that one (clone) of them is well adapted to a certain condition such as a treatment. Metabolic changes for instance can induce cancer therapy resistance. Nutrient starvation may lead to autophagy in cancer cells and concomitant growth arrest—a state called "**dormancy**" eventually resulting in resistance to clinical interventions [56]. The dormancy state is reversible meaning that cancer cells can re-enter hyperactive proliferation once conditions permit.

Reversibility of functional states is an important aspect of tumour plasticity. A prime example is the epithelial-mesenchymal transition (EMT), which cancer cells transiently undergo for metastatic dissemination. Once these cells reached a distant tissue to colonize, they need to reverse the EMT process and perform mesenchymal-epithelial transition (MET). In the case of lung cancer, the full spectrum between EMT and MET phenotype was revealed by single-cell mRNA sequencing [54]. Such analyses hint to a huge challenge in cancer therapy. As tumours evolve and adapt quickly and diversely, it is very hard to treat them.

9.3 Combat Cancer

The fight against cancer begins before its actual occurrence by raising awareness and cancer **prevention**. Early detection is an important indicator of a good prognosis. Regular check-ups have to be performed routinely. **Diagnosis** means the actual detection of a tumour, be it benign or malignant, using technologies like histology (i.e. inspection of the tissue's structure and composition) or PET scans as mentioned above. Besides detection of cancer cells in respective organs, the cancer patients need to be stratified. Stratification means that the type and stage of cancer are defined, which determines an efficient **treatment**.

9.3.1 Current Therapeutic Approaches

There are different ways to treat cancer:

- **Surgery**: The malignant tissue is surgically removed. The biggest problem is that it is hard to ensure that all cancer cells were cut out and nothing is growing back.
- **Chemotherapy**: Dose of chemical drugs like Cisplatin, which targets rapidly dividing cells by binding DNA thereby inhibiting DNA replication. Those drugs are not very specific and affect also other cells that divide quickly such as within hair follicles. This is a reason why chemotherapy patients lose their hair. Other drugs are small molecules and usually kinase inhibitors targeting growth-factor signalling.
- **Irradiation**: High doses of locally confined radiation aim at killing tumour cells. Most challenging is to specifically hit only cancer cells and as little as the possible surrounding tissue, particularly if the target is moving as in breathing lung cancer patients.
- **Immunotherapy**: Engineering mostly the adaptive immune system for the recognition and targeted eradication of cancer cells. A prime example involves the blockade of negative regulators of the immune response, e. g. CTLA-4 [67].

9.3.2 the Future of Personalized Medicine

As molecular networks are part of the aetiology (i.e. origin and cause) of the disease, they can be part of the solution to cure cancer. The entire genome, transcriptome or proteome can now be quantitatively assessed (see Sect. 3.4.2). These "omics" data become more and more available for individual patients even at multiple time points throughout the course of the disease. Single-cell information depicts cancer cell heterogeneity. What needed now, are mathematical formalisms to integrate those big data into innovative, interdisciplinary approaches to offer a new generation of personalized medicine.

Each cancer patient has an individual history of the disease. Powerful machine learning approaches can be utilized to address these in a patient-specific manner

[78]. And therefore, we need you, dear reader. With your expertise in mathematics, computer sciences, physics and engineering, biomedical research will benefit largely from your conceptual input supporting us on our ultimate quest to combat cancer.

Summary

- Cancer is not just one disease, but there are many different types and subtypes depending on when and where tumours are initiated in the body.
- Cancer occurs individually in every patient depending on genetics, lifestyle and environment.
- Despite common hallmarks of cancer, combating it requires a personalized approach of prevention, diagnosis and treatment.

Glossary

antigen Agent that can evoke an immune reaction.

apoptosis Programmed cell death including cell shrinkage, fragmentation and coordinated decay with remaining debris being eaten up by neighbouring cells.

autophagy Cellular self-digestion to remove excessive contents through degradation and recycling.

biopsy Taking cells out of the living organism.

carcinogenesis Process of cancer development induced by a carcinogenic substance.

chaperone Proteins that guide correct folding of other proteins or preventing them from the formation of aggregates.

clone Cell with very similar genetic code and shared ancestry among other clones.

consensus A sequence, e.g. of DNA, which is very similar in different species and thus evolutionary conserved.

cyclin D2 Molecule that is up- and down-regulated during the progression of the cell cycle.

cyclin G2 Molecule that is up- and down-regulated during the progression of the cell cycle.

cytokine Secreted protein that affects nearby cells that carry a cognate cytokine receptor.

dysplasia Dysplastic growth of a tissue results in abnormal cells, for instance, with larger nuclei.

endomembrane system Highly interconnected network of membrane-enclosed organelles including endoplasmatic reticulum, Golgi apparatus, lysosomes, peroxisomes and endosomes of eukaryotic cells.

enzyme A molecular machine that catalyses a certain chemical reaction.

epithelium Thin tissue sheet forming the outer surface of our body and lining inner cavities such as the gut.

gene Section of DNA that is transcribed to an RNA molecule that either exerts a task by itself or is translated to a protein with a specific function.

HeLa Cervix cancer cell line derived from the patient Henrietta Lacks.

© Springer-Verlag GmbH Germany, part of Springer Nature 2022
L. Adlung, *Cell and Molecular Biology for Non-Biologists*,
https://doi.org/10.1007/978-3-662-65357-9

helicase An enzyme that catalyses the unwinding of a helix, e.g. the DNA double helix.

homoeostasis Maintenance of a physiological process under steady-state conditions.

hyperplasia Hyperplastic growth of a tissue results in an overall increase in cell number such that there are more of all cells.

kinase An enzyme catalysing the addition of a phosphate group to its substrate.

lesion Damage or partial loss of tissue resulting from injury or disease.

ligand Molecule recognized by a receptor.

malignancy Presence of a malignant tumour, synonymous for cancer.

metaplasia Almost normally looking cells are found in disparate places.

mitogen Extracellular molecule that stimulates cell-cycle progression and division.

mutation Change in the genetic code deviating from the original version.

Myc Origin: Myelocytomatosis, a transcription factor involved in cell-cycle progression.

necrosis Cell death induced by a cellular injury leading to spillage of cytoplasmic contents to the surroundings.

neoplasia A process leading to the formation of new tissue as cell types are found in disparate places creating a neoplasm.

non-homologous end joining Mechanism of DNA repair.

nucleus Membrane-enclosed compartment within a cell storing and organizing heritable information as DNA.

oncogene A gene that encodes for a protein responsible for uncontrolled growth and division of cells. Overexpression or hyperactivation of those genes often leads to cancer. E.g. Ras.

ortholog Gene version in different species, such as the Ras gene in humans and mice encoding for a protein with the same function but slightly different DNA sequence.

p53 Tumour suppressor, "guardian of the genome" as the main regulator of DNA damage response, aka tumour protein TP53.

pathogen A substance, such as a bacterium or a virus, that can cause a disease.

phosphatase An enzyme catalysing the subtraction of a phosphate group from its substrate.

proliferation Process of repeated cell divisions upon completed cell cycles to increase cell numbers.

protease An enzyme that degrades proteins by cleaving peptide bonds between certain amino acids.

proteasome A cellular machinery that degrades proteins.

pSTAT5 Phosphorylated and thus activated signal transducer and activator of transcription 5.

quiescent Cell stopped in cell cycle and thus not dividing.

Raf Origin: rapidly accelerated fibrosarcoma, family of cancer-related serine/threonine protein kinases.

Ras Origin: Rat sarcoma, small molecule capable of GTP hydrolysis, involved in signal transduction.

receptor Protein structure incorporated into the cellular membrane to bind extracellular ligands and activate intracellular processes.

senescence Cellular ageing and deterioration leading to dysfunction and eventually growth arrest.

SNARE Transmembrane proteins that face the cytoplasmic side of vesicles and target membranes and wind up to catalyse fusion.

syncytium A cell containing multiple nuclei, usually at an early stage of embryonic development.

transcription factor Molecule capable of DNA binding for control of gene expression.

tumour suppressor A cellular factor preventing cells from uncontrolled growth and division thereby blocking tumour formation, e.g. p53, often found mutated or inactive in cancer. ubiquitination Addition of a single or several small molecules of ubiquitin to a protein that is thereby labelled for degradation.

Bibliography

1. Lorenz Adlung. Dependence of E. coli Chemotaxis on CheB Phosphorylation in Silico and in Vivo. *Journal of Unsolved Questions*, 2(1):Articles 1–4, 2012.
2. Lorenz Adlung and Ido Amit. From the Human Cell Atlas to dynamic immune maps in human disease. *Nature Reviews Immunology*, 18(10):597–598, oct 2018.
3. Lorenz Adlung, Sandip Kar, Marie-Christine Wagner, Bin She, Sajib Chakraborty, Jie Bao, Susen Lattermann, Melanie Boerries, Hauke Busch, Patrick Wuchter, Anthony D Ho, Jens Timmer, Marcel Schilling, Thomas Höfer, and Ursula Klingmüller. Protein abundance of AKT and ERK pathway components governs cell type-specific regulation of proliferation. *Molecular Systems Biology*, 13(1):904, jan 2017.
4. Lorenz Adlung, Paul Stapor, Christian Tönsing, Leonard Schmiester, Luisa E. Schwarzmüller, Lena Postawa, Dantong Wang, Jens Timmer, Ursula Klingmüller, Jan Hasenauer, and Marcel Schilling. Cell-to-cell variability in JAK2/STAT5 pathway components and cytoplasmic volumes defines survival threshold in erythroid progenitor cells. *Cell Reports*, 36(6):109507, aug 2021.
5. Shizuo Akira, Satoshi Uematsu, and Osamu Takeuchi. Pathogen recognition and innate immunity. *Cell*, 124(4):783–801, feb 2006.
6. Bruce Alberts. *Essential cell biology*. Garland Science, New York ; London, 4. ed. edition, 2013.
7. Bruce Alberts, Dennis Bray, Karen Hopkin, Alexander D Johnson, Julian Lewis, Martin Raff, Keith Roberts, and Peter Walter. *Essential cell biology*. Garland Science, New York, NY, 4. ed. edition, 2014.
8. L Alexopoulou, A Holt, R Medzhitov, and R Flavell. Recognition of double-stranded RNA and activation of NF-kappaB by Toll-like receptor 3. *Nature*, 413(6857):732–738, 2001.
9. M A Atkinson and E H Leiter. The NOD mouse model of type 1 diabetes: as good as it gets? *Nature medicine*, 5(6):601–4, jun 1999.
10. Julie Bachmann, Andreas Raue, Marcel Schilling, Martin E Böhm, Clemens Kreutz, Daniel Kaschek, Hauke Busch, Norbert Gretz, Wolf D Lehmann, Jens Timmer, and Ursula Klingmüller. Division of labor by dual feedback regulators controls JAK2/STAT5 signaling over broad ligand range. *Molecular Systems Biology*, 7(516):1–15, jul 2011.
11. Basil Bakir, Anna M Chiarella, Jason R Pitarresi, and Anil K Rustgi. EMT, MET, Plasticity, and Tumor Metastasis. *Trends in Cell Biology*, 30(10):764–776, oct 2020.
12. Verena Becker, Marcel Schilling, Julie Bachmann, Ute Baumann, Andreas Raue, Thomas Maiwald, Jens Timmer, and Ursula Klingmüller. Covering a broad dynamic range: information processing at the erythropoietin receptor. *Science (New York, N.Y.)*, 328(5984):1404–8, jun 2010.

© Springer-Verlag GmbH Germany, part of Springer Nature 2022
L. Adlung, *Cell and Molecular Biology for Non-Biologists*,
https://doi.org/10.1007/978-3-662-65357-9

13. Nicolai J Birkbak and Nicholas McGranahan. Cancer Genome Evolutionary Trajectories in Metastasis. *Cancer cell*, 37(1):8–19, jan 2020.

14. Elizabeth H Blackburn, Elissa S Epel, and Jue Lin. Human telomere biology: A contributory and interactive factor in aging, disease risks, and protection. *Science (New York, N.Y.)*, 350(6265):1193–8, dec 2015.

15. Maria A Blasco. Telomeres and human disease: ageing, cancer and beyond. *Nature Reviews Genetics*, 6(8):611–622, aug 2005.

16. Martin Erich Boehm, Lorenz Adlung, Marcel Schilling, Susanne Roth, Ursula Klingmüller, and Wolf Dieter Lehmann. Identification of Isoform-Specific Dynamics in Phosphorylation-Dependent STAT5 Dimerization by Quantitative Mass Spectrometry and Mathematical Modeling. *Journal of proteome research*, 13(12):5685–94, oct 2014.

17. N J Burroughs, B M P M Oliveira, A A Pinto, and M Ferreira. Immune response dynamics. *Mathematical and Computer Modelling*, 53(7-8):1410–1419, apr 2011.

18. Katrin Busch, Kay Klapproth, Melania Barile, Michael Flossdorf, Tim Holland-Letz, Susan M. Schlenner, Michael Reth, Thomas Höfer, and Hans-Reimer Rodewald. Fundamental properties of unperturbed haematopoiesis from stem cells in vivo. *Nature*, 518(7540):542–6, feb 2015.

19. Jayaram Chandrashekar, Mark A Hoon, Nicholas J P Ryba, and Charles S Zuker. The receptors and cells for mammalian taste. *Nature*, 444:288–294, 2006.

20. Qian Cheng and Jiandong Chen. The phenotype of MDM2 auto-degradation after DNA damage is due to epitope masking by phosphorylation. *Cell cycle (Georgetown, Tex.)*, 10(7):1162–6, apr 2011.

21. Yong Cheng, Zhihai Ma, Bong-Hyun Kim, Weisheng Wu, Philip Cayting, Alan P. Boyle, Vasavi Sundaram, Xiaoyun Xing, Nergiz Dogan, Jingjing Li, Ghia Euskirchen, Shin Lin, Yiing Lin, Axel Visel, Trupti Kawli, Xinqiong Yang, Dorrelyn Patacsil, Cheryl A. Keller, Belinda Giardine, Anshul Kundaje, Ting Wang, Len A. Pennacchio, Zhiping Weng, Ross C. Hardison, and Michael P. Snyder. Principles of regulatory information conservation between mouse and human. *Nature*, 515(7527):371–375, nov 2014.

22. Koei Chin, Carlos Ortiz de Solorzano, David Knowles, Arthur Jones, William Chou, Enrique Garcia Rodriguez, Wen-Lin Kuo, Britt-Marie Ljung, Karen Chew, Kenneth Myambo, Monica Miranda, Sheryl Krig, James Garbe, Martha Stampfer, Paul Yaswen, Joe W Gray, and Stephen J Lockett. In situ analyses of genome instability in breast cancer. *Nature Genetics*, 36(9):984–988, sep 2004.

23. V J Cid, I Rodriguez-Escudero, A Andres-Pons, C Roma-Mateo, A Gil, J den Hertog, M Molina, and R Pulido. Assessment of PTEN tumor suppressor activity in nonmammalian models: the year of the yeast. *Oncogene*, 27(41):5431–5442, .

24. Richard Coico and Geoffrey Sunshine. Immunology : a short course. John Wiley & Sons. 7th edition, 2015 (ISBN: 978-1-118-39691-9).

25. Manuel Collado and Manuel Serrano. Senescence in tumours: evidence from mice and humans. *Nature Reviews Cancer*, 10(1):51–57, jan 2010.

26. Galen Andrew Collins and Alfred L. Goldberg. The Logic of the 26S Proteasome. *Cell*, 169(5):792–806, may 2017.

27. Jesse W Cotari, Guillaume Voisinne, Orly Even Dar, Volkan Karabacak, and Gregoire Altan-Bonnet. Cell-to-Cell Variability Analysis Dissects the Plasticity of Signaling of Common gamma Chain Cytokines in T Cells. *Science Signaling*, 6(266):ra17, mar 2013.

28. Thomas Cremer and Marion Cremer. Chromosome territories. *Cold Spring Harbor perspectives in biology*, 2(3):a003889, mar 2010.

29. Jesse R. Dixon, Siddarth Selvaraj, Feng Yue, Audrey Kim, Yan Li, Yin Shen, Ming Hu, Jun S. Liu, and Bing Ren. Topological domains in mammalian genomes identified by analysis of chromatin interactions. *Nature*, 485(7398):376–80, apr 2012.

30. John D'Orazio, Stuart Jarrett, Alexandra Amaro-Ortiz, and Timothy Scott. UV radiation and the skin. *International journal of molecular sciences*, 14(6):12222–48, jun 2013.

31. Kevin K. Duclos, Jesse L. Hendrikse, and Heather A. Jamniczky. Investigating the evolution and development of biological complexity under the framework of epigenetics. *Evolution & Development*, 21(5):276–293, sep 2019.

32. Greg Elgar and Tanya Vavouri. Tuning in to the signals: noncoding sequence conservation in vertebrate genomes. *Trends in genetics : TIG*, 24(7):344–52, jul 2008.

33. Pablo Engel, Laurence Boumsell, Robert Balderas, Armand Bensussan, Valter Gattei, Vaclav Horejsi, Bo-Quan Jin, Fabio Malavasi, Frank Mortari, Reinhard Schwartz-Albiez, Hannes Stockinger, Menno C van Zelm, Heddy Zola, and Georgina Clark. CD Nomenclature 2015: Human Leukocyte Differentiation Antigen Workshops as a Driving Force in Immunology. *Journal of immunology (Baltimore, Md. : 1950)*, 195(10):4555–63, nov 2015.

34. James E Ferrell, Tony Yu-Chen Tsai, and Qiong Yang. Modeling the cell cycle: why do certain circuits oscillate? *Cell*, 144(6):874–85, mar 2011.

35. Michalis Fragkos, Olivier Ganier, Philippe Coulombe, and Marcel Méchali. DNA replication origin activation in space and time. *Nature reviews. Molecular cell biology*, 16(6):360–74, jun 2015.

36. Thomas F Gajewski, Hans Schreiber, and Yang-Xin Fu. Innate and adaptive immune cells in the tumor microenvironment. *Nature Immunology*, 14(10):1014–1022, oct 2013.

37. Chutian Ge, Jian Ye, Ceri Weber, Wei Sun, Haiyan Zhang, Yingjie Zhou, Cheng Cai, Guoying Qian, and Blanche Capel. The histone demethylase KDM6B regulates temperature-dependent sex determination in a turtle species. *Science*, 360(6389):645–648, may 2018.

38. Amir Giladi and Ido Amit. Single-Cell Genomics: A Stepping Stone for Future Immunology Discoveries. *Cell*, 172(1-2):14–21, jan 2018.

39. M B Ginzberg, R Kafri, and M Kirschner. On being the right (cell) size. *Science*, 348(6236):1245075–1245075, may 2015.

40. Sara Goodwin, John D. McPherson, and W. Richard McCombie. Coming of age: ten years of next-generation sequencing technologies. *Nature Reviews Genetics*, 17(6):333–351, jun 2016.

41. Eric D Green, James D Watson, and Francis S Collins. Human Genome Project: Twenty-five years of big biology. *Nature*, 526(7571):29–31, oct 2015.

42. Angela T Hahn, Joshua T Jones, and Tobias Meyer. Quantitative analysis of cell cycle phase durations and PC12 differentiation using fluorescent biosensors. *Cell cycle (Georgetown, Tex.)*, 8(7):1044–52, apr 2009.

43. Douglas Hanahan and Robert A. Weinberg. Hallmarks of cancer: the next generation. *Cell*, 144(5):646–74, mar 2011.

44. Aaron N Hata, Jeffrey A Engelman, and Anthony C Faber. The BCL2 Family: Key Mediators of the Apoptotic Response to Targeted Anticancer Therapeutics. *Cancer discovery*, 5(5):475–87, may 2015.

45. Edith Heard. Dosage compensation in mammals: fine-tuning the expression of the X chromosome. *Genes & Development*, 20(14):1848–1867, jul 2006.

46. Ronald Hoffman, Edward J. Benz, Leslie E. Silberstein, Helen Heslop, Jeffrey Weitz, and John Anastasi. *Hematology: Basic Principles and Practice*. Elsevier Saunders, Philadelphia, PA, 6 edition edition, 2013.

47. Ed Hurt and Martin Beck. Towards understanding nuclear pore complex architecture and dynamics in the age of integrative structural analysis. *Current opinion in cell biology*, 34:31–8, jun 2015.

48. F Jacob and J Monod. Genetic regulatory mechanisms in the synthesis of proteins. *Journal of molecular biology*, 3:318–56, jun 1961.

49. Diego Adhemar Jaitin, Lorenz Adlung, Christoph A Thaiss, Assaf Weiner, Baoguo Li, Hélène Descamps, Patrick Lundgren, Camille Bleriot, Zhaoyuan Liu, Aleksandra Deczkowska, Hadas Keren-Shaul, Eyal David, Niv Zmora, Shai Meron Eldar, Nir Lubezky, Oren Shibolet, David A. Hill, Mitchell A. Lazar, Marco Colonna, Florent Ginhoux, Hagit Shapiro, Eran Elinav, and Ido Amit. Lipid-Associated Macrophages Control Metabolic Homeostasis in a Trem2-Dependent Manner. *Cell*, 178(3):686–698.e14, jul 2019.

50. Guixiang Ji, Yan Long, Yong Zhou, Cong Huang, Aihua Gu, and Xinru Wang. Common variants in mismatch repair genes associated with increased risk of sperm DNA damage and male infertility. *BMC Medicine*, 10(1):49, dec 2012.

51. Xinming Jing, Fengming Yang, Chuchu Shao, Ke Wei, Mengyan Xie, Hua Shen, and Yongqian Shu. Role of hypoxia in cancer therapy by regulating the tumor microenvironment. *Molecular Cancer*, 18(1):157, dec 2019.

52. Russell G Jones and Craig B Thompson. Tumor suppressors and cell metabolism: a recipe for cancer growth. *Genes & Development*, 23(5):537–548, mar 2009.

53. Greg Kabachinski and Thomas U Schwartz. The nuclear pore complex–structure and function at a glance. *Journal of cell science*, 128(3):423–9, feb 2015.

54. Loukia G Karacosta, Benedict Anchang, Nikolaos Ignatiadis, Samuel C Kimmey, Jalen A Benson, Joseph B Shrager, Robert Tibshirani, Sean C Bendall, and Sylvia K Plevritis. Mapping lung cancer epithelial-mesenchymal transition states and trajectories with single-cell resolution. *Nature Communications*, 10(1):5587, dec 2019.

55. Verena A Katzke, Rudolf Kaaks, and Tilman Kühn. Lifestyle and cancer risk. *Cancer journal*, 21(2):104–10, 2015.

56. Candia M Kenific, Andrew Thorburn, and Jayanta Debnath. Autophagy and metastasis: another double-edged sword. *Current Opinion in Cell Biology*, 22(2):241–245, apr 2010.

57. Raj Khanna. Immunology.

58. Daniel C Kirouac, Julio Saez-Rodriguez, Jennifer Swantek, John M Burke, Douglas A Lauffenburger, and Peter K Sorger. Creating and analyzing pathway and protein interaction compendia for modelling signal transduction networks, 2012.

59. H Kitano, A Funahashi, Y Matsuoka, and K Oda. Using process diagrams for the graphical representation of biological networks. *Nat Biotechnol*, 23(8):961–966, .

60. Anja K Klappan, Stefanie Hones, Ioannis Mylonas, and Ansgar Brüning. Proteasome inhibition by quercetin triggers macroautophagy and blocks mTOR activity. *Histochemistry and cell biology*, 137(1):25–36, jan 2012.

61. Edda Klipp, editor. *Systems biology*. Wiley-VCH, Weinheim, 1. repr. edition, 2010.

62. Günter Kramer, Daniel Boehringer, Nenad Ban, and Bernd Bukau. The ribosome as a platform for co-translational processing, folding and targeting of newly synthesized proteins. *Nature structural & molecular biology*, 16(6):589–97, jun 2009.

63. Brent M Kuenzi and Trey Ideker. A census of pathway maps in cancer systems biology. *Nature Reviews Cancer*, 20(4):233–246, apr 2020.

64. U Kummer, L F Olsen, C J Dixon, A K Green, E Bornberg-Bauer, and G Baier. Switching from simple to complex oscillations in calcium signaling. *Biophysical journal*, 79:1188–1195, .

65. D P Lane. p53, guardian of the genome. *Nature*, 358(6381):15–16, jul 1992.

66. Pablo Lara-Gonzalez, Frederick G Westhorpe, and Stephen S Taylor. The spindle assembly checkpoint. *Current biology : CB*, 22(22):R966–80, nov 2012.

67. Dana R Leach, Matthew F Krummel, and James P Allison. Enhancement of Antitumor Immunity by CTLA-4 Blockade. *Science*, 271(5256):1734–1736, mar 1996.

68. Tong Ihn Lee and Richard A Young. Transcriptional regulation and its misregulation in disease. *Cell*, 152(6):1237–51, mar 2013.

69. B Lindemann. Receptors and transduction in taste. *Nature*, 413:219–225, .

70. Jennifer K Litton, Harold J Burstein, and Nicholas C Turner. Molecular Testing in Breast Cancer. *American Society of Clinical Oncology Educational Book*, 39(39):e1–e7, may 2019.

71. David M MacAlpine and Geneviève Almouzni. Chromatin and DNA replication. *Cold Spring Harbor perspectives in biology*, 5(8):a010207, aug 2013.

72. Michela Masetti, Roberta Carriero, Federica Portale, Giulia Marelli, Nicolò Morina, Marta Pandini, Marta Iovino, Bianca Partini, Marco Erreni, Andrea Ponzetta, Elena Magrini, Piergiuseppe Colombo, Grazia Elefante, Federico Simone Colombo, Joke M.M. den Haan, Clelia Peano, Javier Cibella, Alberto Termanini, Paolo Kunderfranco, Jolanda Brummelman, Matthew Wai Heng Chung, Massimo Lazzeri, Rodolfo Hurle, Paolo Casale, Enrico Lugli, Ronald A DePinho, Subhankar Mukhopadhyay, Siamon Gordon, and Diletta Di Mitri. Lipid-loaded tumor-associated macrophages sustain tumor growth and invasiveness in prostate cancer. *Journal of Experimental Medicine*, 219(2), feb 2022.

73. Corbin E Meacham and Sean J Morrison. Tumour heterogeneity and cancer cell plasticity. *Nature*, 501(7467):328–337, sep 2013.
74. S R Menakuru, N J Brown, C A Staton, and M W R Reed. Angiogenesis in pre-malignant conditions. *British Journal of Cancer*, 99(12):1961–1966, dec 2008.
75. D O Morgan. *The Cell Cycle: Principles of Control*. Primers in biology. New Science Press, 2007.
76. Siddhartha Mukherjee. *The Emperor of All Maladies: A Biography of Cancer*. Scribner, 2010.
77. D E Nelson, A E C Ihekwaba, M Elliott, J R Johnson, C A Gibney, B E Foreman, G Nelson, V See, C A Horton, D G Spiller, S W Edwards, H P McDowell, J F Unitt, E Sullivan, R Grimley, N Benson, D Broomhead, D B Kell, and M R H White. Oscillations in NF-kappaB signaling control the dynamics of gene expression. *Science (New York, N.Y.)*, 306:704–708, 2004.
78. Kee Yuan Ngiam and Ing Wei Khor. Big data and machine learning algorithms for health-care delivery. *The Lancet Oncology*, 20(5):e262–e273, may 2019.
79. E A Nigg. Nucleocytoplasmic transport: signals, mechanisms and regulation. *Nature*, 386(6627):779–87, apr 1997.
80. Vassiliki Nikoletopoulou, Maria Markaki, Konstantinos Palikaras, and Nektarios Tavernarakis. Crosstalk between apoptosis, necrosis and autophagy. *Biochimica et Biophysica Acta (BBA) - Molecular Cell Research*, 1833(12):3448–3459, dec 2013.
81. Dominik Niopek, Dirk Benzinger, Julia Roensch, Thomas Draebing, Pierre Wehler, Roland Eils, and Barbara Di Ventura. Engineering light-inducible nuclear localization signals for precise spatiotemporal control of protein dynamics in living cells. *Nature communications*, 5:4404, jan 2014.
82. Dominik Niopek, Pierre Wehler, Julia Roensch, Roland Eils, and Barbara Di Ventura. Optogenetic control of nuclear protein export. *Nature communications*, 7:10624, jan 2016.
83. Béla Novák and John J Tyson. Design principles of biochemical oscillators. *Nature reviews. Molecular cell biology*, 9(12):981–91, dec 2008.
84. Tomoyo Okada, Miguel Lopez-Lago, and Filippo G Giancotti. Merlin/NF-2 mediates contact inhibition of growth by suppressing recruitment of Rac to the plasma membrane. *Journal of Cell Biology*, 171(2):361–371, oct 2005.
85. C. Pan, C. Kumar, S. Bohl, U. Klingmueller, and M. Mann. Comparative Proteomic Phenotyping of Cell Lines and Primary Cells to Assess Preservation of Cell Type-specific Functions. *Molecular & Cellular Proteomics*, 8(3):443–450, oct 2008.
86. Franziska Paul, Ya'ara Arkin, Amir Giladi, Diego Adhemar Jaitin, Ephraim Kenigsberg, Hadas Keren-Shaul, Deborah Winter, David Lara-Astiaso, Meital Gury, Assaf Weiner, Eyal David, Nadav Cohen, Felicia Kathrine Bratt Lauridsen, Simon Haas, Andreas Schlitzer, Alexander Mildner, Florent Ginhoux, Steffen Jung, Andreas Trumpp, Bo Torben Porse, Amos Tanay, and Ido Amit. Transcriptional Heterogeneity and Lineage Commitment in Myeloid Progenitors. *Cell*, 163(7):1663–77, nov 2015.
87. Leïla Perié, Ken R Duffy, Lianne Kok, Rob J de Boer, and Ton N Schumacher. The Branching Point in Erythro-Myeloid Differentiation. *Cell*, 163(7):1655–62, dec 2015.
88. Allison Piovesan, Maria Chiara Pelleri, Francesca Antonaros, Pierluigi Strippoli, Maria Caracausi, and Lorenza Vitale. On the length, weight and GC content of the human genome. *BMC Research Notes*, 12(1):106, dec 2019.
89. D M Prescott. Regulation of cell reproduction. *Cancer research*, 28(9):1815–20, sep 1968.
90. Abe Pressman, Celia Blanco, and Irene A Chen. The RNA World as a Model System to Study the Origin of Life. *Current biology : CB*, 25(19):R953–63, oct 2015.
91. Ruibao Ren. Mechanisms of BCR-ABL in the pathogenesis of chronic myelogenous leukaemia. *Nature Reviews Cancer*, 5(3):172–183, mar 2005.
92. Shimon Sakaguchi, Tomoyuki Yamaguchi, Takashi Nomura, and Masahiro Ono. Regulatory T cells and immune tolerance. *Cell*, 133(5):775–87, may 2008.
93. M Schilling, T Maiwald, S Bohl, M Kollmann, C Kreutz, J Timmer, and U Klingmuller. Computational processing and error reduction strategies for standardized quantitative data in biological networks. *FEBS J*, 272(24):6400–6411, 2005.

94. M Schilling, T Maiwald, S Hengl, D Winter, C Kreutz, W Kolch, W D Lehmann, J Timmer, and U Klingmuller. Theoretical and experimental analysis links isoform-specific ERK signalling to cell fate decisions. *Molecular Systems Biology*, 5:334, 2009.

95. Annette Schneider, Ursula Klingmüller, and Marcel Schilling. Short-term information processing, long-term responses: Insights by mathematical modeling of signal transduction: Early activation dynamics of key signaling mediators can be predictive for cell fate decisions. *BioEssays : news and reviews in molecular, cellular and developmental biology*, 34(7):542–50, jul 2012.

96. Gregg L Semenza. Tumor metabolism: cancer cells give and take lactate. *Journal of Clinical Investigation*, 118(12):3835–7, nov 2008.

97. Ron Sender, Shai Fuchs, and Ron Milo. Revised estimates for the number of human and bacteria cells in the body. *bioRxiv*, jan 2016.

98. Liat Shenhav and David Zeevi. Resource conservation manifests in the genetic code. *Science*, 370(6517):683–687, nov 2020.

99. Jennifer Sills, Nirupa Galagedera, Jyoti Mishra, Meryem Ayas, Patrick Kobina Arthur, Islam Mosa, P. William Hughes, Lei Jiao, Sam Allon, Kara Kuntz-Melcavage, M. Romina Schiaffino, Poonam C. Singh, Swati Negi, Rivca E. Hildebrandt, Lorenz Adlung, Collet Dandara, and Liqiang Zhong. Global collaboration. *Science*, 346(6205):47–49, oct 2014.

100. Ryan J Taft, Michael Pheasant, and John S Mattick. The relationship between non-protein-coding DNA and eukaryotic complexity. *BioEssays : news and reviews in molecular, cellular and developmental biology*, 29(3):288–99, mar 2007.

101. J M Thevelein, R Geladé, I Holsbeeks, O Lagatie, Y Popova, F Rolland, F Stolz, S Van de Velde, P Van Dijck, P Vandormael, A Van Nuland, K Van Roey, G Van Zeebroeck, and B Yan. Nutrient sensing systems for rapid activation of the protein kinase A pathway in yeast. *Biochemical Society transactions*, 33:253–256, 2005.

102. Tsvi Tlusty. A colorful origin for the genetic code: information theory, statistical mechanics and the emergence of molecular codes. *Physics of life reviews*, 7(3):362–76, sep 2010.

103. Henry Tat Kwong Tse, Westbrook McConnell Weaver, and Dino Di Carlo. Increased asymmetric and multi-daughter cell division in mechanically confined microenvironments. *PloS one*, 7(6):e38986, jan 2012.

104. R Y Tsien. The green fluorescent protein. *Annual review of biochemistry*, 67:509–44, jan 1998.

105. Mohan E Tulapurkar, Aparna Ramarathnam, Jeffrey D Hasday, and Ishwar S Singh. Bacterial lipopolysaccharide augments febrile-range hyperthermia-induced heat shock protein 70 expression and extracellular release in human THP1 cells. *PloS one*, 10(2):e0118010, jan 2015.

106. John J Tyson, Katherine C Chen, and Bela Novak. Sniffers, buzzers, toggles and blinkers: dynamics of regulatory and signaling pathways in the cell. *Current opinion in cell biology*, 15(2):221–31, apr 2003.

107. Béla Völgyi, Feng Pan, David L Paul, Jack T Wang, Andrew D Huberman, and Stewart A Bloomfield. Gap junctions are essential for generating the correlated spike activity of neighboring retinal ganglion cells. *PloS one*, 8(7):e69426, jan 2013.

108. James D Watson. Recombinant DNA : genes and genomes - a short course. W.H. Freeman and Company, 3rd edition, 2007 (ISBN: 0-7167-2866-4).

109. Robert Weinberg. *The Biology of Cancer*. John Wiley & Sons Ltd, West Sussex, England, 2 edition, jan 2007.

110. Robin A Weiss. A perspective on the early days of RAS research. *Cancer metastasis reviews*, 39(4):1023–1028, 2020.

111. E John Wherry and Makoto Kurachi. Molecular and cellular insights into T cell exhaustion. *Nature Reviews Immunology*, 15(8):486–499, aug 2015.

112. Eileen White and Robert S DiPaola. The Double-Edged Sword of Autophagy Modulation in Cancer. *Clinical Cancer Research*, 15(17):5308–5316, sep 2009.

113. Jeffrey B Wyckoff, Yarong Wang, Elaine Y Lin, Jiu-feng Li, Sumanta Goswami, E Richard Stanley, Jeffrey E Segall, Jeffrey W Pollard, and John Condeelis. Direct Visualization of

Macrophage-Assisted Tumor Cell Intravasation in Mammary Tumors. *Cancer Research*, 67(6):2649–2656, mar 2007.

114. Guang Yao, Tae Jun Lee, Seiichi Mori, Joseph R Nevins, and Lingchong You. A bistable Rb-E2F switch underlies the restriction point. *Nature cell biology*, 10(4):476–82, apr 2008.

115. Ying-Kiat Zee, James P B O'Connor, Geoff J M Parker, Alan Jackson, Andrew R Clamp, M Ben Taylor, Noel W Clarke, and Gordon C Jayson. Imaging angiogenesis of genitourinary tumors. *Nature Reviews Urology*, 7(2):69–82, feb 2010.

116. L Zhang and C Wang. F-box protein Skp2: a novel transcriptional target of E2F. *Oncogene*, 25(18):2615–27, apr 2006.

Printed in the United States
by Baker & Taylor Publisher Services